A HISTORY OF THE

ZINC SMELTING INDUSTRY

IN BRITAIN

Published on the occasion of the half centenary of
The Imperial Smelting Corporation Group

A HISTORY OF THE
ZINC SMELTING INDUSTRY
IN BRITAIN

BY

E. J. COCKS
Formerly Scholar of Christ's College, Cambridge

AND

B. WALTERS
Formerly of Sidney Sussex College, Cambridge,
and the University College of Swansea

GEORGE G. HARRAP & CO. LTD
London · Toronto · Wellington · Sydney

First published in Great Britain 1968
by GEORGE G. HARRAP & CO. LTD
182 High Holborn, London, W.C.1

DESIGNED AND PRODUCED BY CHARLES ROSNER & ASSOCIATES
LIMITED, LONDON AND PRINTED AT THE CURWEN PRESS

CONTENTS

PREFACE

THIS IS THE HISTORY of a basic industry which the British Government stimulated into large-scale activity to meet urgent national need in two World Wars but left to fight for its own international existence when peace returned. It has been written in celebration of the fiftieth anniversary of the first beginnings of what is now Imperial Smelting Corporation Limited, the Group which has been, since 1933, the sole producer of primary zinc in Britain. The authors hope that it will be of interest alike to the economic historian, the scientist, and the connoisseur of management and finance expertise. To past and present employees of the Corporation itself and of its parent Company since 1962, the Rio Tinto-Zinc Corporation, it may seem disappointingly lacking in detail about people and events they remember. To this the authors can only reply that, as far as they know, this is the first time that the history of this particular industry in this country has been written and that, in any case, to write a true history of this industry means also writing a history of the numerous useful by-products which zinc smelting has added to the national product. It seemed desirable, therefore, in the time available, to compress essential facts into one volume rather than to collect interesting detail for many. If this volume stimulates others to provide this detail in later books and articles the authors will be the first to rejoice. The young science called Industrial Archaeology is providing ever increasing scope for the publication of information on the struggles and achievements of our forerunners in industry. Coal, iron, steel, textiles and railways have appeared to monopolize the attention of industrial historians hitherto and a great deal more information on the development of the non-ferrous metals and chemical industries would help towards correcting the rather distorted picture that the average text book gives of our continuing Industrial Revolution.

E.J.C.
B.W.

ACKNOWLEDGEMENTS

THE BULK OF the material used for this book comes from the primary evidence of Board records and files, which, through the commendable zeal of past Company and private Secretaries, have been maintained with hardly a single gap since 1929. For the period before that date Minute Books exist but otherwise the evidence is sketchy and a great deal has had to be filled in from the Public Records Office, the archives of the Registrar of Companies, financial accounts, and legal documents. As most of these sources are not readily available to the public, extensive use has been made of quotation to illustrate the text. A list of some of the publications to which reference has been made is appended.

In a book of this nature on a new subject and with an international spread of involved technical and financial detail, the authors could not have completed the task without the valuable help very willingly given by a number of friends and colleagues, whose help is most gratefully acknowledged.

Mr. S. W. K. Morgan, A.R.S.M., B.SC., who has led Imperial Smelting's research effort since 1935 and is widely known in world metallurgical circles as the principal inventor of the Imperial Smelting Process, has contributed much of the scientific information in this book and a unique account of the origins of the invention.

Mr. D. S. Burwood, A.R.S.M., B.SC., who has served the group since 1925 and was for sometime Deputy General Manager, has, with his wide technical knowledge and experience and meticulous regard for accuracy, been a tower of strength throughout. Mr. Emlyn Evans, B.SC., formerly of Swansea University and for many years Works Manager of the historic Swansea Vale Smelting Works, has saved the authors many months of work by placing unreservedly at their disposal the results of years of spare-time research into the early history of zinc in Asia, Europe and Britain.

The authors are privileged also to have received the help and advice of Lord Chandos, whose remarkable career of public service has included twenty-five most active years in the seats of power of the international non-ferrous metals industry, and of the late Lord Baillieu, whose long connection with the international zinc industry needs no introduction, as well as the benefit of the vast experience of the world's metal markets since the 1914–18 War of Mr. A. M. Baer former Chairman of the Rio Tinto-Zinc Corporation, and of Mr. William Mure, C.B.E., Deputy Chairman of British Metal Corporation.

Information is also gratefully acknowledged from Metallgesellschaft, The British Metal Corporation, Mr. J. Askew, Secretary of British Titan Products, Mr. R. Bounds, Secretary of Fison's, Mr. Cecil Fry of Burma Corporation, and

Mr. Geoffrey Blainey of Melbourne University who, by kind permission of Mrs. W. S. Robinson, has placed the British section of the late W. S. Robinson's papers at the disposal of the authors pending the publication of W. S. Robinson's Memoirs.

Shortage of space makes it impossible to give a complete list of colleagues past and present who have given help and advice, except to say that the bulk of the text has been perused by Mr. M. I. Freeman (present Chairman of Imperial Smelting), Lord Kirkwood, Mr. W. P. Harris, Mr. A. V. Broad, Mr. E. J. A. Fry, Mr. L. T. Evans, Mr. E. H. V. Jorey, Mr. A. B. Pitcher, Mr. D. M. G. Sneddon and Mr. H. V. Casson. Mr. G. E. Flack has provided some of the early figures for Appendix II, and Mr. R. S. B. Ames, former Chief Accountant of Imperial Smelting, has checked and advised on most of the financial detail.

A final word of thanks is deservedly due to Pamela Bowl, who typed the whole of the first draft of this book from the authors' obscure handwriting and a large portion of the subsequent drafts in addition to compiling the Index, and to the Publicity Department of the Rio Tinto-Zinc Corporation whose encouragement and help have been readily available throughout three laborious years of preparation.

As regards the text, the word 'Limited' has been omitted from names of Limited Companies to reduce verbiage except where there is a possibility of confusion arising. 'Britain' has been used to include zinc smelting in Wales and save the extra words involved in 'Great Britain' or 'The United Kingdom'.

Philippus Aureolus Theophrastus Paracelsus.

CHAPTER ONE

Early History of Zinc Smelting

ZINC THE DRAB NEWCOMER AMONG METALS—BRONZE AND BRASS
ZINC OXIDE—METALLIC ZINC IN THE FAR EAST
METALLURGY AND ALCHEMY IN INDIA—'ELIXIR VITAE'
DEVELOPMENT OF ZINC SMELTING IN INDIA
ALCHEMY AND ZINC IN CHINA
SMELTING IN CHINA—IMPORT OF ZINC FROM THE EAST
INTO EUROPE—ZINC AND CALAMINE—THE EUROPEAN ALCHEMISTS
AND ZINC—'CADMIA'—ZINC MINERALS—WILLIAM CHAMPION OF
BRISTOL, PIONEER OF THE BRITISH ZINC INDUSTRY
CHAMPION'S PATENTS, PROCESS AND WORKS—HIS OUTPUT
OF ZINC—HIS FALL AND BANKRUPTCY—BEGINNINGS OF ZINC
SMELTING IN THE SWANSEA AREA—IMPORTANT DEVELOPMENTS
ON THE CONTINENT OF EUROPE

ZINC IS NOT a glamorous metal. No one has yet written a waltz about it as Franz Lehar did about 'Gold and Silver'. It does not adorn the shelves of antique shops like copper kettles and warming pans. It is not flashy and familiar like tin, or ancient and venerable like lead. Iron and steel have long been known as basic ingredients of modern industry and aluminium as the symbol of the lighter contemporary outlook. Zinc, when its uses are known at all, is usually associated in popular imagery with ugly corrugated iron roofs and dustbins of the pre-plastic era, in spite of the publicity work of the Zinc Development Association since the war.

This is not surprising. Zinc has been recognized and isolated as a separate metal only in comparatively recent times and, in Britain, has received the exclusive name of 'zinc' instead of 'spelter' only since the thirties of the present century. Although it has been produced in Britain intermittently since the mid-eighteenth century, the necessity for organizing its production as a mass-producing basic industry was not appreciated by the Government until 1914, even though it had been the main ingredient of brass for many centuries and vital for the protection of iron and steel from corrosion by the damp British climate since the beginnings of large-scale industry in these islands.

Copper was known in its metallic state prior to recorded history, tin was known about 3000 B.C., and lead was obtained in the metallic state about 2000 B.C. Brass, an alloy of zinc and copper, was known a short period before the dawn of the Christian era. The numerous references to 'brass' over a millennium earlier in the books of Moses, particularly in Exodus, actually refer to bronze, the copper-tin alloy, which has given its name to a famous archaeological 'Age'.

Brass making was, in fact, the earliest known outlet for zinc and still remains one of its four most important uses, but it was not until the mid-nineteenth century that the modern method of producing brass by alloying actual metals succeeded the centuries-old method of 'cementation' which was in essence alloying with zinc vapour. In this process granulated copper or copper shot was mixed with zinc oxide in the form of calcined calamine and with charcoal (or—later—anthracite) and heated in a sealed crucible. The zinc vapour from the reduction of the oxide combined with the copper to give brass.

The zinc oxide used for this purpose was known early in the Christian era as there is evidence that it was recovered in a fairly pure form from specially constructed dust chambers during copper smelting operations, some zinc mineral being present in the copper ore. Later it was produced in such places as Cyprus and Persia by 'sublimation', a process described by Marco Polo as he witnessed it at the state controlled works in Persia, and this type of zinc oxide, known by several names according to its degree of purity, was used sometimes for medical purposes as it is to this day.

The numerous outlets for zinc will become apparent as this history proceeds but its other important uses in the modern world, principally for galvanizing, die casting and rolling, have developed only with the Industrial Revolution while some of its former and more exotic uses have disappeared and little is known about them.

The early history of metallic zinc is meagre and the evidence indicates that it originated in the Far East. Probably the first specimen was produced in India on what would now be described as a laboratory scale. In these early ages metallurgy and alchemy were inextricably mixed and, in fact, remained so in Europe until after the Middle Ages.

Alchemy flourished in India during the period A.D. 700–1300 and the basis of the art was essentially the search for a material that would confer supernatural powers upon the taker and prolong life.

The earlier alchemists were only to a lesser degree concerned with the dream of their successors of being able to convert the base metals into gold. The Hindu scriptures of this period reveal that they distilled zinc oxide with various fats, oils and other carbonaceous materials such as wool and that the zinc vapour, which was thus distilled off in small quantities, was cooled, condensed in water and then mixed with other materials to provide the 'Elixir Vitae'—the supposed source of wonderful powers.

Development of the actual art of zinc smelting was slow. There is evidence that, in the fourteenth century, small earthenware retorts about 18 in. long by 2 in. internal diameter were used for distillation in Mewar State and were packed with a charge consisting of zinc oxide and a reducing agent. Zinc smelting virtually did not develop beyond this stage in India.

Communications between India and China existed at least from the first century A.D. and there are records of Buddhist monks from China visiting India and writing accounts of their impressions. There was sea traffic between the countries from early in the Christian era and missions were established in China, but there is no evidence of whether any knowledge about zinc passed between the two countries in this way. Certainly, alchemy and zinc in India and in China appear to have followed similar courses.

In China alchemy may have started before the Christian era. One of the alchemists' aims was, again, the search for the medicine that could prolong life —'Chin Tau'—but there is no evidence that zinc was one of the ingredients tried in these experiments.

Tracing the history of zinc in China is extremely difficult as the name of 'zinc' has been repeatedly changed. Zinc was used as coinage in China in A.D. 1402, so that it must have been produced earlier in sufficient quantities and on a commercial scale, however small, to meet this demand. In a Sino-Japanese encyclopaedia published in 1713 the various names for zinc were listed and among them appears 'Totan'—hence the suggestion that this word was derived from a Sanskrit or Tamil word indicating the possible origin of zinc as being in India.

In the early days smelting in China was carried out in clay crucibles heated by a slow fire. These were filled with a charge consisting of zinc ore and charcoal, sealed with a cover, and fitted with necks in which the zinc vapour cooled and condensed. After this, clay retorts were tried as in India, but, despite the fact that a crude furnace setting was later introduced, progress in China was incredibly slow and some of these retort furnaces were still in use in the present century. Zinc deposits are widely spread in China but generally the veins were only sufficient to support small village industries. The industry never developed beyond the primitive stage.

During the seventeenth century much zinc was imported from the East into the Continent of Europe by the Portuguese fleets, the metal being then known generally, but not universally, as 'Tutenag', and its principal use was probably in ornamental castings such as small statuary for indoor and outdoor decoration.

It is recounted that in 1611 one of the Portuguese ships carrying zinc was captured by the Dutch, who named the substance 'Speauter' which is presumably the same word as 'piauter' meaning 'pewter'. 'Speauter' is the origin of the word 'Spelter' which has been the accepted word for unwrought zinc until recent times. Pewter is an alloy consisting principally of four parts of tin to one part of lead and this possible evolution of the word 'Spelter' is interesting as reflecting the belief widely prevalent until the mid-eighteenth century that zinc was a debased form of some other metal instead of a separate metal with distinctive properties.

The source and method of production of imported zinc were unknown to the importers but it was during the epoch of the sixteenth and seventeenth centuries that dabblers in this field passed to the final recognition that zinc was the same as 'tutenag', and could be made from calamine. At the start of the period the mumbo-jumbo of alchemy was paramount—as, for example, in the idea that as there were seven planets there could only be seven metals. Calamine was the earth thought by them to be capable of transmuting copper into gold (although it must have been quite obvious to the alchemists themselves that the product was far lighter in weight than genuine gold) and there was also a strange metal which kept on appearing in unexpected places and was so like tin, silver and bismuth, but which no one connected with imported tutenag.

For example, in Germany for many centuries lead smelting had been carried out in small shaft furnaces in the Harz region. The lead ore of the area contained varying amounts of zinc and in 1617 Lohneyss, a German chemist, wrote that when lead smelting was carried out there formed in the crevices of the walls a metal called zinc or 'conterfeht' and that when the walls were scraped the metal droplets ran into a trough placed to receive them. He claimed that this metal was not of much commercial value and that the workmen would only collect the 'zinc' if they were promised a bribe as an incentive.

While the separate metal zinc had been recognized by this time many confused it with bismuth although this metal was not isolated until nearly a century later. For instance, Lohneyss also stated that 'alchemists have a great desire for this zinc or bismuth'.

There are scattered references to some of these alchemists and dabblers in zinc in the chronicles of this period. Basil Valentine, a shadowy figure but probably a Benedictine monk, who was also an amateur goldsmith with medical learnings and lived until about 1540, is reputed to have been familiar with the metal itself. 'Zinc' probably derives from a German word meaning jagged or pointed, which could refer either to crystals of calamine or of condensed zinc, but more probably to the former.

Paracelsus (1493–1541), a German-Swiss professor of medicine at Basle, refers to zinc as a 'Bastard of copper', fusible but not malleable, and 'bastard' because the metals corresponding with the seven planets were already known. Georgius Agricola (1495–1555), whose real name was George Bauer, was a devotee of mineralogy and mining and his great work *De Re Metallica* remains even now a metallurgical classic. In this book his 'contrefey' probably refers to zinc, i.e. an imitation of tin or silver, or even of gold when combined with copper. In two other books he mentions zinc by name and refers to an important source of zinc ore in the Salzburg area.

Andreas Libavius (1545–1610), another medical man and alchemist, refers to zinc in a number of his many books on alchemy, but when he received a sample of imported zinc from Holland he obviously did not know what it was. He knew about 'cadmia' which was impure zinc oxide scraped off walls of furnaces smelting lead and silver and was useful for transmuting copper into 'gold' (i.e. brass). To add to the general confusion the word cadmia had no connection with the metal cadmium which, although similar to and occurring with zinc, was unknown at the time.

Even at this stage no one seems to have thought of producing zinc directly by reduction with carbon of either cadmia or the mineral calamine, and it is probable that one of the first to do so was William Champion of Bristol (1710–94) an account of whose achievement will be given later in this chapter.

Confusion was also arising over zinc minerals. At this time the only source of zinc was calamine (i.e. zinc carbonate)*, a reddish, often earthy, material which readily lost its combined carbon dioxide, the resulting oxide being easily reduced to the metal. This was also the earth of the alchemists. The principal lead ore was, and still is, galena (lead sulphide), a black lustrous mineral which is sometimes rich in silver. The early German miners were somewhat puzzled and annoyed to find that some ore, which was very like galena, produced no lead or silver when smelted. This they christened 'Blende' meaning 'Blind' and it is now known as zinc blende, zinc sulphide or 'sphalerite', which is derived from the Greek word meaning 'treacherous': the mineral was also known as mock lead and false galena. The more picturesque name 'Black Jack', while less derogatory, may also spring from the same association of ideas or be merely descriptive.

It is easy to see how this confusion arose. Zinc blende, when free from iron, is a light straw-coloured mineral, which the miners could perhaps have recognized but when it contains appreciable amounts of dissolved iron sulphide it assumes the black colour of 'black jack' which allowed it to be confused with galena.

The first person in Europe to describe the reduction of zinc from calamine was Marggraf, the German chemist, in 1746, but his fellow countryman, Henckel, claimed that he had produced zinc from calamine in 1721 and had concealed the method. Despite the fact that Marggraf's methods could undoubtedly have been developed on a larger scale, no effort was apparently made on the Continent and the first full-scale plant was built in England by William Champion of Bristol about the year 1743.

It is impossible to understand the appearance of this pioneer of the British zinc industry in what might otherwise be regarded as still predominantly an agricultural epoch without taking into account the fact that the Champion family had been prominent in the industrial life of Bristol for some two centuries, their principal interests including the famous and now coveted Bristol pottery. William Champion and his brother John were the sons of Nehemiah

*This is the current British terminology. In the U.S.A. 'calamine' refers to zinc silicate and zinc carbonate is known as smithsonite.

EARLY PORTUGUESE SAILING SHIP

ZINC SMELTING IN CHINA

Champion, the third Nehemiah in the line, who had been engaged in the brass and copper trade and had helped to found Bristol Brass Wire Company in 1702.

The Champion family were Quakers who, in addition to living their quiet and philanthropic lives, were 'famous for their knack of prospering in honestly conducted business' although the family feud recounted later in this chapter casts some doubt on this idyllic description. They were an inventive family, their patents ranging from metallurgy to devices for hatching chickens.

In 1738 William Champion was granted a patent, No. 564, in which the wording is extremely obscure but which indicates that it was at last realized that zinc blende or 'black jack' was a potential source of zinc.

The patent was couched in a mysterious manner, referring to the treatment of sulphurous minerals to obtain 'metallick sulphur' and in it Champion claimed that 'he had with great labour and expense found a method or invention for the reducing of sulphurous blendes minerals and mineral into a body of metallick sulphur known by the name or names of spelter or Tuteneg'. A second patent, taken out twenty years later in 1758 by his brother and rival John Champion, was more specific and describes how blende must be freed from other minerals and stone, finely crushed, and sufficiently calcined before being reduced with charcoal to give spelter or be made into brass.

Considerable argument has ensued over the years as to whether William Champion was the originator of zinc smelting in Britain and as to how he obtained his information.

It has been claimed, for example, that a Dr. Lane succeeded in obtaining zinc in 1717 at his copper and lead works near Swansea but there is no evidence whatsoever to substantiate this claim. Neither is there any evidence to support the statement that a Dr. Isaac Lawson was instrumental in introducing zinc smelting to Britain.

It has also been claimed that Champion learned the secret of making zinc from a sailor who had been in China. This is not credible for foreigners at this period were not allowed to travel inland in China and the small Chinese smelters were far inland from the trading posts. Also when Champion applied, in 1750, for a continuation of his patent beyond the period originally granted he said that he had travelled widely on the Continent of Europe to study manufactories but he never claimed that he had any special knowledge of China. Furthermore, zinc smelting was not introduced on the Continent until the end of the eighteenth century and the method of smelting set up by Champion was totally different from that practised in China.

Whatever the answer to this problem, Champion did, in about 1740, set out to produce zinc from calamine by 'distillation per descensum', and started his

B

plant, for the production of copper, brass and zinc, at Warmley in the Parish of Siston, near Bristol, in 1743. Calamine was available from the Mendip Hills and coal from nearby Kingswood. Few visitors were allowed in the works and the first account was not written until his patent had expired. A Swedish metallurgist, who visited the works in 1754, later reported that there were fifteen copper furnaces, fifteen brass furnaces and four spelter furnaces in operation.

Doubtless Champion's knowledge of furnaces used for glass- and brass-making enabled him to improve on earlier furnace design and his zinc process came to be known as the English process.

In this process the charge, consisting of calcined calamine, was placed inside a large crucible some 3 feet high and 21 inches wide at the top. Each crucible was charged through a hole in the top cover which was then sealed. There were six to eight crucibles per furnace and each crucible had a 4-inch diameter hole in the bottom to which an iron pipe abutted. When the temperature of the charge reached distillation point the zinc vapours passed downwards and cooled in the iron pipes and the condensed zinc would drop into a pan of water placed underneath the end of the pipe. The recovered zinc was melted down and cast into convenient sized ingots or plates.

The furnace was usually octagonal at the base, which consisted of a cave, built of massive masonry pillars, containing the fire and receivers to catch condensed zinc. The pillars supported the circular smelting floor on which the pots were placed, these being covered by a dome in which holes over each pot led to a conical chimney, the whole structure being some 40 feet high.

Over a period of some six years Champion's total output was only about 200 tons of zinc which was claimed to be of better quality than that imported from the Far East and was well received by the Birmingham and Wolverhampton merchants. The purpose for which it was used is not known but it may have been for the whitening of pewter. It was only a sideline to copper and brass production.

It appears that Champion's various furnaces were set in rows and that to the north of the works he built a thirteen acre lake which supplied water for driving his mills. He also built cottages for his workmen and a Dutch style house for himself surrounded by many acres of meadowland. Although few traces of Champion's Works remain, some of the cottages and traces of his extensive landscaping of the area are still to be seen.

Champion was badly hit by the import of cheap foreign zinc, which forced the price down from £260 to £48 per ton. This and his ambitious attempts to rival his brother's concern, the Bristol Brass Wire Company, eventually led to his bankruptcy and the take-over of his works by his rivals.

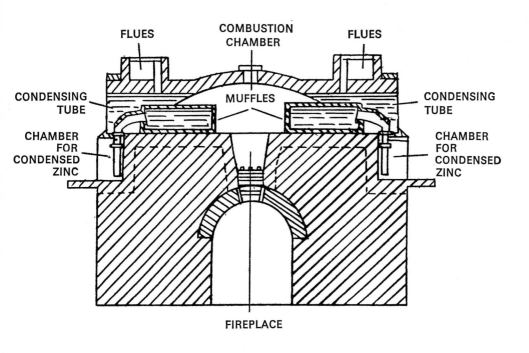

SILESIAN SMELTING FURNACE

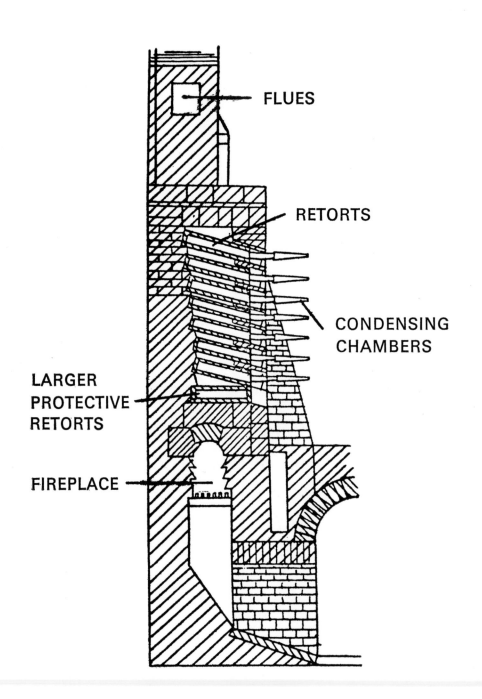

FLUES

RETORTS

CONDENSING CHAMBERS

LARGER PROTECTIVE RETORTS

FIREPLACE

EARLIER BELGIAN COAL FIRED SMELTING FURNACE
(Section along retorts)

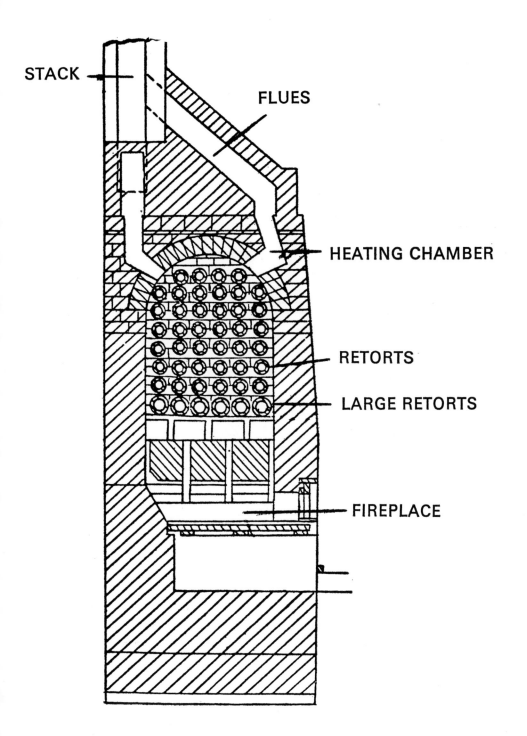

STACK

FLUES

HEATING CHAMBER

RETORTS

LARGE RETORTS

FIREPLACE

EARLIER BELGIAN COAL FIRED SMELTING FURNACE
(Section across retorts)

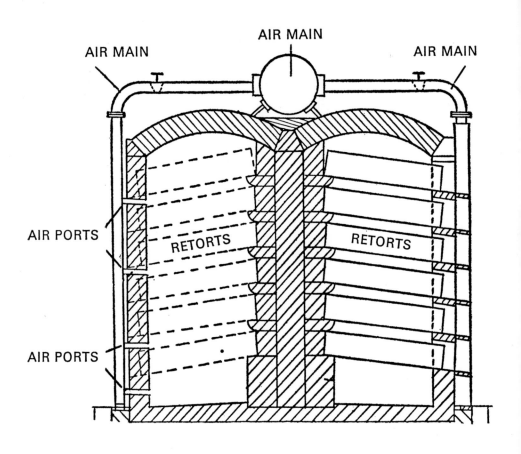

LATER (HEGELER) TYPE OF PRODUCER GAS FIRED BELGIAN FURNACE

He failed in 1750 to obtain a continuation of his patent and his quarrel with his old firm, the Bristol Brass Wire Company, became more bitter. He instituted an expansion programme in the early 1760's and bought into stock considerable quantities of local coals with the sole object of denying coals to his rivals, who retaliated by dropping the price of copper and brass and began to produce zinc themselves. The Warmley company were soon in difficulties especially as they lacked working capital. In 1768 they sought royal assent for a Charter of Incorporation so that the public could subscribe for shares but this petition was strongly opposed by other brass manufacturers and pin makers. A second petition was more successful but the company were by then in a precarious financial position, particularly as, out of their £300,000 of capital, over £200,000 had been borrowed at high interest rates. The Bristol company continued its policy of price cutting and in turn cornered the coal market. The Warmley company went into liquidation and Champion was gazetted a bankrupt in 1769. All the assets of the company were purchased by the Bristol Brass Wire Company and Emerson, former Manager at Warmley Works, built a copper and zinc works at Hanham. The later history of this works is unknown, although it was still prospering at the end of the century according to a contemporary account of Bristol in 1794 by Matthews.

In 1794 also Champion died at the age of eighty-four, but the works which he had founded ceased operation only in 1880, by which time zinc smelting had become firmly established in the Swansea area.

The first beginnings of zinc smelting near Swansea are also traceable in this period, although they were not on the scale of the efforts made by Champion and his contemporaries in Bristol.

Chauncey Townsend, a rich alderman of the City of London, who held coal mining leases in the area East of Swansea, obtained, in 1754, a lease to build a lead and zinc smelter at Upper Bank and, in 1757, to build a copper smelter at the adjacent site of Middle Bank on the east bank of the river Tawe.

Conditions for the establishment of a zinc industry in this area were most propitious. As the English process used some twenty-five tons of coal per ton of zinc produced, transport charges were an important factor. Adequate supplies of semi-bituminous coal and anthracite were available locally and the river was navigable for ships of moderate burden for some miles. Sufficient supplies of low refractory clay for the process could be obtained locally while the more refractory and better quality clays could be obtained from Stourbridge. In addition there were inherent metallurgical skills available as copper and lead smelters had existed in the area for many years.

However, the Upper Bank Works did not survive long as a zinc smelter. In 1775 it was converted into a copper smelter and zinc smelting ceased as an industry in Swansea Vale for over half a century, most probably because of little demand for zinc.

Important developments were also taking place on the Continent of Europe during this period. Since the fifteenth century calcined calamine from the vast deposits at Aachen had been sold to the brass manufacturers in Belgium and in 1806 mining concessions were granted to the Abbé Dony of Liège on the condition that he experimented in the production of metallic zinc from calamine. His experimental work was ultimately successful and the Belgian zinc smelting industry, which grew to be one of the largest in the world during the nineteenth century, started thus in the Liège area. The Belgian method evolved by Dony was different from the English process and, instead of crucibles, consisted of a number of retorts set horizontally in rows.

Large calamine deposits also existed in Upper Silesia and here again the calamine had been used over a considerable period for the manufacture of brass. Then a certain Johann Ruhberg visited England and saw the English process in operation. On his return to Silesia in 1798 he built a zinc smelter at Wessola using a modified glass furnace technique and ten large retorts per furnace.

At this time also a zinc smelting process was introduced at Dollach in Hungary which in many respects resembled the process used in the Far East and consisted of a series of retorts set vertically in a furnace setting. These retorts, with the open end downwards, fitted into holes in the furnace floor which led to a common condensing chamber. This process was short-lived but the Belgian and Silesian industries expanded rapidly and were destined to provide the foundation on which the modern Continental zinc industry of the nineteenth and twentieth centuries was built.

Indeed the rise of these industries to a dominating position in zinc smelting in the second half of the nineteenth century and the failure of the older local industries in India, China and Britain to compete for the status of modern mass-producer is the main theme of the next chapter.

CHAPTER TWO

The Zinc Industry
in the
Nineteenth Century

BRITAIN'S LEAD IN EUROPEAN INDUSTRIAL PROGRESS AFTER
THE NAPOLEONIC WAR NOT SHARED BY ITS ZINC INDUSTRY
LARGE MODERN ZINC INDUSTRIES BUILT UP BY THE UNITED
STATES AND GERMANY IN THE NINETEENTH CENTURY
PROLIFERATION OF SMALL ZINC ENTERPRISES IN BRITAIN DURING
THE INDUSTRIAL REVOLUTION—WEAKNESS OF THE INDUSTRY
DISCOVERY OF A VAST NEW SOURCE OF RAW MATERIALS FOR
LEAD AND ZINC PRODUCTION AT BROKEN HILL,
AUSTRALIA, IN 1883—BULK OF AUSTRALIAN OUTPUT TO
GERMANY AND BELGIUM FOR SMELTING
BRITAIN'S DEPENDENCE ON GERMANY FOR MOST OF HER ZINC
METAL SUPPLIES UP TO 1914
1914—THE MOMENT OF TRUTH FOR THE BRITISH ZINC INDUSTRY
BRITAIN'S DEPENDENCE ON EUROPE FOR ZINC BEFORE 1914
A REFLECTION OF HER YEARS OF PEACE

THE NINETEENTH CENTURY saw the transition from the interesting experiments described in the previous chapter, designed vaguely to produce a metal, the uses of which, except for brass-making, were not yet fully understood, to the large-scale zinc industry of the present day.

It is unsafe to generalize about trends but invention and enterprise seem largely to have deserted zinc smelting during the last quarter of the eighteenth century, after the bankruptcy of William Champion. The one or two scattered developments on the Continent which have been described seem to have taken place in spite of rather than because of the Napoleonic War. In fact, unlike the two major wars of this century, the Napoleonic War appears to have had a retarding rather than a stimulating effect on science and heavy industry and Napoleon himself fought his battles with equipment and metals not vastly different from those used by Marlborough a century before.

After the war, however, Britain, which had remained untouched by invasion, very soon took the lead in European industrial progress and was quick to take advantage of her own inventions in steel, steam, railways, textiles, coal and iron. Her three eventual rivals, France, Germany and the United States, were all handicapped by various forms of political disunity or weakness until the second half of the century. Unfortunately, the lead so readily taken by Britain in other directions did not create the urge to build up a large-scale zinc industry. It has been estimated that corrosion destroys one-fifth* of the world production of ferrous metals annually and at the present time more than 100,000 tons of zinc are used every year in Britain for galvanizing steel against corrosion, but the British zinc industry did not expand in step with the steel industry in the nineteenth century. Its moment of truth did not arrive until 1914 and the outbreak of the First World War. After the American Civil War and the unification of Germany under the leadership of Prussia, both of which momentous events took place in the 'sixties of the nineteenth century, the United States and Germany rapidly built up large modern zinc industries. Britain left its own small-scale zinc works to look after themselves, even after the discovery of a vast deposit of lead and zinc minerals at Broken Hill, Australia, in 1883. The bulk of the Australian output went to Germany for smelting in Germany and Belgium and Britain drew most of its zinc metal supplies from the Continent right up to 1914 when the country found itself faced suddenly with the need to create its own large-scale zinc industry or to depend on the United States of America for its zinc supplies for war purposes instead.

*Vide Zinc Development Association Technical Notes No. 1. 'Hot Dip Galvanising'. Revised April 1959.

Eventually, in 1917, it created its own large-scale industry under the name of The National Smelting Company. The reasons for the delay and the reasons why Britain, after Champion, lost its technical lead of at least half a century over its rivals in zinc production and then stood by in the mid-nineteenth century to allow them to gain a further half-century's lead in building up a commercial scale industry, lie deep in the history of the industry itself and of the European and Dominion politics of the pre-1914 era. An attempt will be made to explain the salient features of the situation in this chapter but it is not possible in a book of this size to give to the amazing industrial and technical expansion of the U.S.A. zinc industry and of Metallgesellschaft in this period the space that they deserve.

Throughout the nineteenth century there was no sign of a revival of the zinc industry in the Bristol area but a few small zinc smelters were commissioned in Sheffield, Birmingham and North Wales. Zinc ore came to Sheffield from Alston Moor and to Birmingham from the Mendips while North Wales had its own zinc deposits. The Tindale Zinc Extraction Company was established near Carlisle and Brands Pure Spelter Ltd., in Ayrshire.

The natural advantages of Swansea as an area for the establishment of a zinc industry were mentioned in the previous chapter and early in the nineteenth century the industry revived there using the English process.

A small plant was built at Loughor (now Llwchwr) between Swansea and Llanelly (now Llanelli) and another near Maesteg. In 1835 Evan John, a native of Llansamlet, also built a small smelter containing three English type furnaces and two calciners alongside the Smith's canal which extended from the river Tawe at Upper Bank to Llansamlet. John's capital, however, was limited and, in 1840, after failing to form the Cambrian Spelter Company, he advertised the works for sale. It remained idle for a considerable time until taken over by Dillwyn, the member of Parliament for Swansea, under whose ownership it gradually expanded. Dillwyn replaced the English process with the Silesian process and ultimately with the Belgian process which brought a considerable saving in coal consumption and increased output.

The proliferation of small zinc enterprises, each depending on the enterprise of one man or a small group of men and not always soundly financed, was typical of this mid-nineteenth century period of the 'Industrial Revolution'. The demand for zinc had increased rapidly over the years. Direct alloying to produce brass was followed by the discovery in 1805 of the conditions in which zinc could be rolled into sheets and in 1836 the discovery of how to galvanize iron, culminating in the patents of Sorel and Crawford, provided further impetus for the demand for zinc. Galvanized steel sheets became increasingly popular for export to the rapidly developing regions of Australia and America, and to other parts of the world.

For some years also George Muntz had been carrying out experiments on the production and rolling of copper/zinc alloys and in 1832 obtained a patent for the alloy Muntz metal with a preferred composition of 60 per cent copper/40 per cent zinc and resistant to the corrosive action of seawater. The primary object of making this alloy was to replace copper for sheathing the wooden hulls of ships and Vivian & Sons, the principal copper smelters in the Swansea area, were quick to see the danger of this new alloy to their copper trade. After litigation they obtained a licence to manufacture Muntz metal themselves and in 1841 purchased the defunct Ynis copper works in Morriston and converted it into a zinc smelter. In 1868 Vivians also purchased the old Fforest copper works, which was adjacent to the Ynis Works, and converted this also into a further zinc smelter. Owing to a shortage of raw materials, however, the Ynis works closed down in 1875 and the Fforest works thereafter became known as the Morriston Spelter Works which ultimately ceased operations in 1926.

With one famous exception—The Swansea Vale Spelter Company—the other small smelters established in the Swansea area in this epoch eventually met a similar fate.

In 1866 Shackleton & Ford, manufacturers of railway stock at St. Thomas in Swansea built a smelter adjoining their wagon works but shortly afterwards sold out to the Swansea Zinc Company who, within a few years, went into liquidation. The works were taken over by the Richardson family's Crown Zinc Company and renamed 'The Crown Works'. In 1883 they sold out to the English Crown Spelter Company which operated the works until their closure in 1930. English Crown started in 1883 with initial issued capital of 21,000 shares of £5 each (£3. 10s. 0d. paid) i.e. £73,500 and by 1913 had property and stocks worth £135,000. Its net profit in that year was just over £4,000—a meagre return on capital—but rose to over £33,000 by 1920. However, by 1928 it was losing heavily and wound up in 1931 when the assets were acquired by Imperial Smelting Corporation.

Pascoe Grenfell & Sons, who had been operating the Upper Bank Works, mentioned in the previous chapter, as a copper smelter, changed over to zinc smelting in 1868 and thereafter the Upper Bank Works continued as a zinc smelter until its closure in 1924.

In 1873 a group of local men started the Villiers Spelter Works which ceased working for some years between 1900–14 and, after changing hands many times, finally closed down in 1924.

In 1876 the only custom smelter, which is still surviving in the area, was commissioned and is now the Swansea Vale Works of Imperial Smelting. The Company was formed as The Swansea Vale Spelter Company Limited and all the shareholders were local worthies. The initial objects were the production of zinc, lead, tinned plates and black plates and the initial issued capital of this public company was 159 shares of £50 each (£28 paid) i.e. £4,452. By 1910 the issued capital had grown to £16,720 but there was a charge of £5,000 on the assets.

These early years of the Swansea Vale Company coincided with significant changes in the technical outlook of the British zinc industry.

By 1876, when the Company was formed, the English and Silesian processes had been abandoned by the local smelters in favour of the so-called Belgian method of smelting and the introduction of this method and the gradual super-session of oxide by sulphide ores confronted all these small firms with technical problems which they were incapable of solving without more capital and foreign assistance.

Calamine had been the zinc ore mainly used up to this period but with the increased demand domestic supplies became inadequate. Ore began to be imported from Italy and Spain and zinc blende had become more widely used. Adequate supplies of blende were obtainable from Cardiganshire, the North Wales area, and the Isle of Man and the number of mines in Wales was considerable.

Then, in the last quarter of the nineteenth century, the vast Broken Hill lead/zinc deposits were discovered and exploited and many attempts were made in the Swansea area to treat what were then termed the complex lead-zinc ores. At that time concentration of the valuable minerals in the ore was carried out by methods depending on differences in specific gravity but the lead and zinc minerals were so intimately mixed that only partial separation was possible by this method. None of the patented methods were success-ful and it was not until the flotation method was developed, mainly in Australia, whereby the lead and zinc portions in the ore could be separated much more efficiently, that abundant supplies of a zinc concentrate were assured. The Welsh industry, however, was not in a position to use abundant supplies.

When calamine supplies began to dwindle the use of blende was considered a necessary evil and the use of the sulphur dioxide from the roasting process for the manufacturing of sulphuric acid was belated. Hence, in the earlier years the vegetation in the Swansea Valley area was devastated by discarded sulphur fumes from both zinc and copper works. The introduction of any measure to improve conditions was carried out only reluctantly if it did not directly improve outputs.

To a considerable extent also the insignificance of the industry in the British economy in later Victorian times was due to an uncharacteristic resistance to change and a reluctance to accept any innovation to improve either the process or conditions. The basic principles of the smelting process in Wales had, over the years, remained virtually static, although many changes to improve outputs, efficiencies and working conditions had been introduced on the Continent. Belgians were engaged as plant superintendents in one or two instances and, in general, Welsh smelters appeared content to follow developments introduced elsewhere. For example, despite the obvious advantages of this invention, many years elapsed before the first approach was made to continental manufacturers leading to the ultimate installation of hydraulic presses for the manufacture of retorts.

The industry also was conservative in a market which was steadily expanding. The Welsh smelters were ready to join the European smelters in a cartel to maintain prices by limiting production but they also united to keep the price of domestic ores as low as possible. Under normal conditions they were unable to compete with the Silesian and Belgian producers and they passed through a still leaner period at the end of the century when the use of natural gas was introduced among the smelters of the American mid-west. The consequent decrease in production costs permitted the Americans for the first time to export zinc to Europe and undersell the European producers.

It is not surprising therefore to find that in the decade before 1914 two of the largest of the Welsh smelters described earlier in this chapter, the Upper Bank Works and The Swansea Vale Spelter Company, had fallen under the control of two German metal houses, Merton and Hirsch respectively.

The Metallgesellschaft story is a remarkable one and it is important to consider it here as the success of Metallgesellschaft in building up the zinc industry of the pre-1914 German Empire may be regarded as largely responsible for stimulating the British Empire reaction which launched the British, the Australian, and, eventually, the Canadian zinc industries after the Anglo-German war broke out. Also, the atmosphere introduced into the metal industry by the German cartel mentality, which was in its heyday from 1870 to 1914, was to influence the world metal industry as a whole right up to 1939 and to linger on afterwards in spite of anti-Trust legislation in various forms and countries.

Metallgesellschaft came into being on 17 May 1881, when Wilhelm Merton, one of the first businessmen ever to indulge in scientific market research, converted the family business in metal ware, exchange, commission, and forwarding into an Aktiengesellschaft (Joint Stock Company) with a capital of two million Marks for 'trade in and manufacture of metals and oxides'. The atmosphere of Imperial expansion in Germany at the time was favourable to rapid growth particularly as this enterprise, in its origins earlier in the century, was fundamentally a banking concern. For an industry to spring from and continue in association with a parent Bank was and still is very usual continental practice. Metallgesellschaft's commercial activities, which later on included the buying and selling of ores, led to manifold investments in mining, smelting and metal working industries. Before 1914 these companies already included the Braubach lead smelters (1896), Metallurgische Gesellschaft (Lurgi) (founded 1897), Berzelius, Berg und Metallbank (1906) 'Metallhutte Kall' G.m.b.H. (1907), which produced blister copper and copper alloys, Schweizerische Gesellschaft für Metallwerke (1910) associations with several other companies including Hoboken (Belgium) Penoles (Mexico) and American Metal Co. Ltd. A world-wide network of representatives was built up including the London firm of Henry R. Merton & Co. who, with their parent company, owned a substantial interest in Pascoe, Grenfell & Sons of Swansea (i.e. Upper Bank Works mentioned previously). Another German firm, Aron Hirsch & Sohn of Halberstadt, gained control of the majority of the shares of The Swansea Vale Spelter Company at some date between 1906 and 1910, as appears in the Annual Returns of the time, but there is no evidence that the four relatives who made up this concern had any direct connection with Metallgesellschaft. According to an early document (c. 1916) very kindly given to the authors by Metallgesellschaft, Aron Hirsch & Sohn's refining interests at that time consisted of a works on the Belgian-French frontier and a small refinery at Hamburg. They were, however, joined with Metallgesellschaft and with Beer, Sondheimer & Co. of Frankfurt, whose metal interests lay principally in the Unterweser Company and in Belgium, in a consortium established in Melbourne under the name of the Australian Metal Company for the purchase of Broken Hill zinc ores on the spot in Australia.

There is no room to digress in this book into the stirring tale of the discovery of the Broken Hill lead/zinc field in Australia in 1883 especially as it has already been told in several excellent books. Suffice it to say that, by the turn of the century, the problem of separating zinc ore from the residue dumps of the Broken Hill lead/silver mines had been solved—Metallgesellschaft and Zinc Corporation's stories differ as to how this was done—and in 1905 The Zinc Corporation was incorporated in Victoria to exploit the enormous quantities of zinc sulphide concentrates which thus became available.

It seems that they were not the first or last Company to produce zinc concentrates from dumps, and later from mines as competition between companies on the Broken Hill field to sell concentrates to the Americans in the early years of the 1914–18 War ruined the price obtainable until the 'Zinc Producers Association' was formed in Australia in 1916.

In the decade or so before the war, however, the evidence shows clearly that the Germans, through the Australian Metal Company, dominated the zinc concentrates market in Australia. Metallgesellschaft state that, in 1913, of 433,100 tons of Australian zinc concentrates mined, 134,750 tons were destined for Germany, 234,500 for Belgium, 29,100 tons for France, 25,050 tons for Holland and only 9,700 tons for Britain.

These figures are a reflection of the extent to which Britain had failed to follow the trend in Europe in the second half of the nineteenth century. Between 1880 and 1913 the world output of zinc metal rose from 234,000 to 985,235 tons of which 67 per cent was produced in Europe and 32 per cent in the U.S.A. During this period zinc production in Germany rose from an annual average of 91,965 to 241,070 tons, in the U.S.A. from 22,320 to 286,000 tons but in Britain from only 30,000 to 62,000 tons. By 1913 only one-tenth of the zinc produced in Britain was from indigenous raw material. The rest was produced from imported foreign ore.

Imports of zinc metal into Britain had also risen steeply as the needs of British industry outstripped the capacity of the home industry to supply it with zinc. By 1907 imports amounted to 88,000 tons and had risen to 145,000 tons by 1913 of which 64,000 tons came from Germany and 53,000 tons from Belgium through the agency of Metallgesellschaft houses. Imports from the U.S.A. whose zinc industry underwent a similar expansion in the same epoch and which in 1907 ousted Germany from pride of place as the world's largest producer, amounted to less than 5,000 tons in 1913. The great modern Canadian industry was largely a product of the post-1918 era.

Looking back on Britain's dependence on Europe for zinc before 1914 it is tempting to charge those in authority at the time with lack of foresight. It must be remembered however that, apart from remote Colonial adventures, Britain had not been involved in a large-scale continental war since the Crimean War half a century before. In spite of the young Kaiser's love of military grandeur, the diplomatic impasse between the Triple Entente and the Austro-German Axis, and Colonial and naval rivalry, recent research has suggested that the idea that all this might lead to war with the nation ruled by Queen Victoria's great-nephew was not seriously entertained by any except a small minority in the first decade of the twentieth century. There was none of

TINSMITH

MANUFACTURE OF SHEET IRON

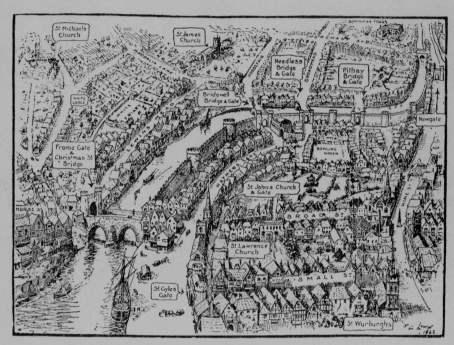

BRISTOL IN THE SEVENTEENTH CENTURY

SWANSEA CASTLE AND HARBOUR

the uncomfortable feeling generally prevalent in the 'thirties that sooner or later the only way to contain the mounting evil personified in Hitler and Mussolini was to go to war.

In this calm Edwardian era before 1914 the non-ferrous metals business was truly international as was shown, for example, by the great friendship existing between W. S. Robinson, at that time Managing Director of the lead smelting works at Port Pirie owned by Broken Hill Associated Smelters, and the German Richard Merton of Henry R. Merton & Co. The idea that Australia and Britain might have to build modern zinc plants to smelt Australian sulphide concentrates, instead of leaving it to Germany and Belgium to do this, nowhere appears in The Zinc Corporation's or any other traceable records before 1914.

C

CHAPTER THREE

Why the Modern British Industry Began

THE IMPORTANCE OF ZINC IN WARTIME—1914 DEPRIVES THE
AUSTRALIAN ZINC MINES OF THEIR MAIN CUSTOMER AND
BRITAIN OF GERMAN AND BELGIAN ZINC—RAPID INCREASE OF
U.S.A. IMPORTS—FEVERISH EFFORTS TO EXPAND THE
BRITISH INDUSTRY—SWANSEA VALE WORKS PASSES FROM THE
GERMANS TO TILDEN SMITH—FORMATION OF THE AUSTRALIAN
ZINC PRODUCERS ASSOCIATION—BIRTH OF THE IMPERIAL SCHEME
FOR ZINC SMELTING—ADVENT OF W. S. ROBINSON
WILLIAM HUGHES SELLS THE COALITION GOVERNMENT
A ZINC CONCENTRATES CONTRACT AND THE AVONMOUTH IDEA
TILDEN SMITH OUTBIDS THE AUSTRALIANS
FOR THE AVONMOUTH VENTURE
RESENTMENT FELT BY BRITISH ZINC MINING INTERESTS

THE IMPORTANCE OF zinc and its by-product, sulphuric acid, in wartime was outlined to the shareholders of Imperial Smelting Corporation by the Chairman, John Govett, in his Annual Statement for 1945:

The main peace-time products of the Imperial Smelting Corporation—zinc and sulphuric acid—have a very wide application in munitions and essential war products. To mention only a few cases, there is zinc as brass in cartridge and shell cases, zinc in die castings for shell fuses, tanks and aircraft components, zinc as dust for smoke screens, zinc as oxide and sulphide for pigments, while sulphuric acid is required for explosives, fertilisers and numerous essential chemicals. . . . Zinc furnaces were extended . . . and a peak production of 75,650 tons of zinc was achieved in the year ending June 30th, 1942.

In contrast, nearly thirty years before, on 27 June 1916, his father, F. A. Govett, of the famous stockbroking firm of that name and Chairman of The Zinc Corporation, had made the following statement to the shareholders in London. As it describes an important link in this history it is quoted in full:

The zinc concentrates before the (1914–18) war were sold under a long-term contract to German smelters, they being the only available purchasers at the time when this business was first started at Broken Hill, and consequently the price of spelter might go to a thousand pounds a ton, and we could not produce or sell a ton of spelter as we have no smelter of our own. . . . On the outbreak of war we found ourselves with our financial resources tied up, and with our wealth unrealisable and, therefore, we could not go on producing and piling up zinc concentrates which then we could not sell.

We immediately tackled the question of erecting smelters for ourselves in England, and within a month from the outbreak of war in August, 1914, we were round at the Board of Trade with a definite proposal to the Government on lines almost identical with the present Imperial scheme. If the Government had then grasped the importance of the situation and seen fit seriously to entertain our offer, we should now (i.e. 1916) actually be producing spelter here in England, but, of course, they did not realise the necessities until too late.

War broke out on 4 August 1914. The National Smelting Company was incorporated on 12 April 1917. The intervening two-and-a-half years in the history of zinc in Britain provide a fascinating story of improvisation and personal intrigue in the presence of the national emergency. Perhaps this was typical of an England which had no idea how to organize for a major war and allowed 'business as usual' to dominate its policy in the first two years of the war until the munitions crisis helped to topple the Asquith Government.

Urgent pressures were building up around zinc both in Britain and Australia. The main problem in Britain was to find another source of supply to replace Germany and Belgium as suppliers of the bulk of the nation's rapidly growing demand for zinc for munitions. Manufacturers turned, naturally, to the U.S.A.

zinc smelters. An article in the *Mining Journal* of 5 February 1916, states that U.S.A. zinc smelters increased production by at least 300 per cent in 1915 mainly in order to replace Germany and Belgium as suppliers of the bulk of Britain's rising zinc requirements. Annual production of zinc metal in the U.S.A., which amounted to 18 tons in 1858, to 56,250 tons in 1890, and to 132,140 tons in 1902, had risen to 286,000 tons by 1913, and to a peak of 598,000 tons by 1917. Among the principal U.S.A. zinc smelters before 1913 were New Jersey Zinc Company, American Smelting and Refining and American Zinc, Lead and Smelting Co., and production was almost entirely from indigenous ore. Again, space does not permit a detailed account of the growth of the U.S.A. zinc industry to be given in this book but one result of all this effort was an increase of imports of U.S.A. zinc into Britain from 4,670 tons in 1913 to 35,000 tons in 1914 and to a peak of 51,000 tons in 1917. This, it will be noted, was considerably less than the total of German and Belgian imports before the war and the explanation is that total British imports of zinc slumped from 145,000 tons in 1913 to a low point of 53,000 tons in 1916 with a modest increase in the last two years of the war. Clearly German U-Boats and shipping difficulties must have had a lot to do with this but the shortage led to a sharp price rise which made zinc very expensive for non-military uses. The annual average price of zinc per ton in the five years before 1914 was about £24. In 1915 and 1916 it had gone up to £66·7 and £69·5 respectively but these averages conceal wide fluctuations between £28 and £115 in the first two war years compared with a variation between 'high' and 'low' not wider than £5 in the years 1909–13. There is no clear evidence but the *Mining Journal*'s statement of 5 February 1916 that the shortage and high price of zinc had destroyed the main peace-time outlet, the galvanizing trade, appears most likely to be accurate. Clearly the bulk of the zinc available was being used for brass manufacture.

Apart from imports, Britain also endeavoured to extract a little more production from its own antiquated zinc industry, the condition of which at this period has already been described. According to summarized data published in 1929 by the U.S. Department of Commerce, Bureau of Mines, no great success followed this effort. Annual average production for the years 1901–5 was 41,204 tons, for 1906–10 56,114 tons, for 1911–15 56,184 tons, and for 1916–20 40,700 tons. A copy of a contract is extant, dated 29 August 1916, between the Minister of Munitions and The Swansea Vale Spelter Company. Under this, in return for extending the works to an annual production capacity of at least 10,000 tons of zinc metal a year (with provision for further extension to 15,000 tons) instead of its 'pre-war standard of output . . . agreed at five thousand tons per annum', various concessions were to be allowed by the Treasury as regards extent of depreciation charged and liability to excess profits tax. The opportunity was quickly taken by the new owners of the Company, about whom much more appears later in this book, to modernize and expand the whole of this works which was, at that date, already forty years old. 'C' plant was built, consisting of eight Belgian type distillation furnaces, and the four old Welsh type furnaces in 'B' plant were replaced by four furnaces of the same Belgian type. The production capacity of the works was thus trebled and a year later the existence of these new Belgian type horizontal furnaces in strength at Swansea influenced the decision to adopt the same process in the proposed works at Avonmouth. The rebuilding work was carried out mainly by Belgian refugees led by M. Van Guelk and, in the early stages, the furnaces burnt down as fast as they were erected owing to faulty bricks. After the initial problems had been overcome, however, Swansea Vale Works, through the strenuous efforts of the men and the management, became the largest producer of zinc metal in Britain until 1934 when the new vertical retort plant started up at Avonmouth.

It is not known whether any of the other small zinc concerns still operating in the Swansea Valley in 1914 also received Government encouragement to expand at this time. None of them survived the post-war economic troubles and the great depression of 1929–31 and their records can no longer be traced in detail.

The Swansea Vale Spelter Company had undergone a change of ownership on the outbreak of the war, which was later to assume great significance in the history of the British zinc industry. In 1914, 802 out of 1,000 shares were owned by the four Hirsch relations and Dr. Emil Hirsch was one of the Directors. This was the same firm that was one of the partners with Metallgesellschaft and Beer, Sondheimer in the Australian Metal Co., mentioned in the previous chapter. By July 1915, the Board of Trade consented to the Hirsch shares being transferred to a Mr. Philip Smith 'having regard to the provisions of . . . the Trading with the Enemy Amendment Act'. Later in 1915 the name of R. Tilden Smith appears as signatory to a letter to the Company outlining a proposed capital reorganization which would arise 'to give effect to the arrangement between myself and the Messrs. Hirsch contained in my contract with them of the 22nd July'. Briefly, as a result of this capital reorganization the Hirsch relations, in spite of the Trading with the Enemy Act, finished up with a holding of £100,000 non-voting Preference Shares in payment for their previous Ordinary Shares and their loan to the Company, and Tilden Smith (500 shares) and the Share Guarantee Trust (1,500 shares), in which he held the controlling interest, finished up as the sole owners of the issued Ordinary Share Capital. Three famous names in the later history of the British zinc industry, R. Tilden Smith, P. Marmion, and H. J. Enthoven appear in the Annual Returns as 'added' as Directors between 1915 and 1916. They joined G. S. and P. W. Smith and J. B. Turnbull who had been Directors with the Hirsch relations before the war.

A great deal appears about Tilden Smith in the next chapter but it must be admitted that, although a lot of incidental facts are known about his actions as a business magnate of First World War vintage and about his Directorships, not much is really known about where he came from, how his fortunes grew, or how he eventually faded away from the industrial scene in the 'twenties.

He appears to have been a colourful character and is shown in the Directory of Directors of 1917 as a director of Burma Corporation, Channel Collieries Trust and Copper Pit Collieries. He had been a Director of Burma Mines, then a highly speculative venture, from 1908 and was certainly a colliery owner, as the early official documents of The National Smelting Company Limited show. Lord Chandos in his memoirs mentions him as a financier and industrialist of great imagination and vision but grandiose ideas, who had largely been financed by Lloyds Bank during the war. W. S. Robinson once described him as 'flamboyant, florid and apparently flourishing'.

His next coup in the zinc world was over the establishment of a zinc smelting works at Avonmouth with the aid of a Government loan of £500,000. The details of this will be given in the next chapter but the events leading up to it were taking place in England and Australia while Swansea Works was being expanded under its new management and an attempt to knit together the scattered threads of evidence is made in the rest of this chapter.

As will have been noted from the speech of F. A. Govett quoted early in this chapter The Zinc Corporation made some sort of offer, of which details are not now traceable, to erect a zinc smelting works in Britain on the outbreak of the war but this offer appears to have been declined. F. A. Govett's public speeches to his shareholders in these years are most informative and from them can be pieced together the situation which confronted the zinc concentrates producers on the Broken Hill field in the first two years of the war and which formed the background to subsequent events in Britain. Deliveries to the Continent of Europe were no longer possible, considerable stocks were in hand, and the zinc concentrator was shut down to prevent further increase of stocks. 'Spot' sales were arranged by several of the Broken Hill companies to U.S.A.

KING EDWARD VII VISITING BRITISH INDUSTRY (*c.*1908)

AN EARLY ROTARY FURNACE AT ORR'S ZINC WHITE (1919)

BROKEN HILL—WHERE IT BEGAN

GENERAL VIEW OF THE OPEN CUT

AVONMOUTH WORKS IN 1932

SWANSEA VALE WORKS IN 1932

LLOYD GEORGE   SIR ROBERT HORNE

W. S. ROBINSON

The whole of this deal for which, it was alleged, Runciman, McKenna and Bonar Law of Asquith's coalition administration were mainly responsible on the British side, was debated in the House of Commons in a debate on the Estimates in 1922, when Lloyd George was Prime Minister. It came in for some pungent criticism and became a hardy annual both for the House and the Committee of Public Accounts throughout the 'twenties. It was attacked not only by the representatives of the dying English lead/zinc mines, who found themselves undercut by cheaper Australian concentrates, but also by those who thought that the Coalition Government had either been excessively naive or excessively devious over the whole negotiation. By 1922 the zinc price had dropped to £30 a ton whereas it was nearly £70 when the contract was negotiated so that the Government were faced with a tremendous loss, even if they could dispose of the concentrates. The loss was estimated by the Parliamentary Secretary concerned at £500,000 by as early as 1922. The final loss to the Government was shown by the Comptroller and Auditor-General's report on the Trading Accounts (Board of Trade) concerned to have been £8,374,000. This is mentioned in the proceedings before the Select Committee of Public Accounts dated 3 March 1932. The contract was summed up in the Committee's second Report in 1922 although again the financial terms were not stated:

An agreement was concluded in 1917 as regards concentrates and spelter with the Zinc Producers' Proprietary Association (Limited) Australia. This agreement operates until the 30th June, 1930, and is divided into three periods as under:

1. From the 1st January, 1918 to the 30th June, 1921.
2. From the 1st July, 1921 to the 30th June, 1925.
3. From the 1st July, 1925 to the 30th June, 1930.

The annual quantity of concentrates the Government may be required to take is fixed by the agreement at 250,000 tons per annum in the first period, and 300,000 tons per annum for the second and third periods. The prices are fixed for the first two periods and a formula laid down for regulating prices in the third period. There is also an agreement giving the Zinc Producers' Association the right to 'put' 45,000 tons of spelter annually with the Government at ruling market price.

The History of the Ministry of Munitions supplies the missing information on price which both the Minister in Parliament and the Select Committee appear to have been anxious to suppress. The fixed price for the first two periods of the concentrates contract was ninety shillings a ton of concentrates subject to minor conditional variations and for the third period—from 1 July 1925—it varied in accordance with a formula based on the London Metal Exchange price.

The spelter commitment was part of the effort to provide adequate supplies of refined zinc of 99·9 per cent purity for brass and cartridge metal during the war. In Britain the refining plants of Brunner Mond & Co., Chance & Hunt and The New Delaville Company were also encouraged to make a minor contribution to this demand while Stewart & Lloyds were persuaded by the Government to produce refined zinc from hard spelter* by a process invented by Mr. Guy Fricker of Fricker's Metal Company and licensed by Fricker's to the Government as they had no room to use it on their Luton site.

The Government managed to negotiate out of this spelter commitment in the Australian contract in 1922.

Apart from persuading the Government to make this contract, William Hughes, on his visit in 1916, also appears to have been successful in imbuing them with some enthusiasm for the Empire zinc smelting scheme. With the chronic shortage of zinc in the country in 1916 and after the previous fruitless persuasion from F. A. Govett they probably were not difficult to convince that something must be done to increase zinc production in Britain, even if only to make use of some of the concentrates they had bought. About six to eight months later came the incorporation of The National Smelting Company Ltd. and the beginning of work to build a zinc smelter on the Avonmouth site; but the negotiations between the Government, the Tilden Smith interest, and the Australian producer interests have left behind inadequate evidence of what actually went on and of the motives of those involved.

According to F. A. Govett's published statements The Zinc Corporation were not prepared to undertake the responsibility of erecting the proposed smelting works unless they had adequate guarantees of protection, mainly against the threat of a rapid revival of the German zinc industry after the war stifling the infant British industry. F. A. Govett's blunt words on this point, in his speech to Zinc Corporation shareholders of 16 July 1916, are worth quoting in full as they are redolent of the commercial and political atmosphere of the time and prophetic of British Government dealings with the zinc industry up to the present day:

The idea is good; the Imperial sentiment is good; but two points had first to be assured. . . . It is our duty to safeguard the commercial end, leaving consideration of political expediency to Ministers and politicians, and that is the line we have taken, for first, without some sliding protective tariff or a sliding bounty, to proceed with the scheme of building Imperial smelters would be just wasting money, for the industry would be killed by foreign competition. The possibility of the erection of British or Australian smelters on any important scale simply depends on whether the

---

* 'Hard spelter' is zinc contaminated with iron from galvanizing.

Imperial Government can be brought to face the fact that the industry to be established might be protected by bounty and by preferential tariff. It is possible Mr. Hughes may succeed in this Herculean task, but you are aware of the convention of the Coalition (i.e. Lloyd George's new Government) to shelve such inconvenient party questions which involve decisions as to how they will meet German competition in trade when war is done.

The second point was that

while I had no belief in the willingness of the Imperial Government to admit this thin end of the protection wedge, even with that granted if we were not prepared to do the work ourselves and provide the money ourselves I foresaw the greatest difficulties in obtaining reasonable terms from enterprising financiers proposing to erect British smelters for our concentrates, for zinc smelting in itself is not an industry of large profits, except in times of high prices for spelter.

Later on F. A. Govett said quite specifically that failing a co-operative scheme between the Empire producers to erect these smelters, Zinc Corporation were not prepared to proceed on their own unless the Government guaranteed a bounty which maintained zinc metal at the 'living wage' of £23 a ton. Presumably the guarantee was not given as he later reported about 'a new group willing on terms to undertake the English portion of the scheme, leaving our companies to handle the Australian end'. There is no evidence that this was a reference to Tilden Smith's group but it is a permissible assumption that his was the 'new group'.

W. S. Robinson's account of these years, given in his private papers compiled many years after the event, is rather different and mentions a trio consisting of Tilden Smith, Andrew Weir and John Higgins. Andrew Weir was a shipping magnate who later became Lord Inverforth. He was also one of those who put up the money to enable Lloyd George to achieve his ambition of acquiring control of the *Daily Chronicle*. John (later Sir John) Higgins was a chemist who had acquired quite a large smelting works at Dry Creek, near Adelaide, which failed when German interests moved into Australia. This left him with a bitter distaste for anything German. Later he became Weir's personal representative and also honorary metallurgical adviser to the Commonwealth Government. This trio had some other scheme in mind for smelting Australian concentrates in Britain and W. S. Robinson says that it was only by promising to frame an Empire scheme quickly that he managed to talk William Hughes out of supporting Tilden Smith's scheme. The interview lasted an hour and a half and took place at the Cecil Hotel in the Strand in April 1916. W. S. Robinson suggests that Tilden Smith and Weir were attempting 'to get hold of what they thought to be a good thing'. If so, they must have been disappointed at the outcome. It is described in the next two chapters.

There is nothing else in the scattered papers of W. S. Robinson which are accessible, or in the papers of The Zinc Corporation at this time to suggest that, after the preliminary rebuff in 1914, they were anxious to be entrusted with the Avonmouth scheme by the Government in 1916. The terms of their offer and of Tilden Smith's rival offer, both of which required a substantial amount of Government finance, are set out in Volume 7 of the History of the Ministry of Munitions but there appears to be very little difference in the two offers except that Tilden Smith was obviously prepared to agree to a larger measure of Government control than were either The Zinc Corporation or the Zinc Producers Association. The 'History' gives no clear statement of why Tilden Smith's offer was preferred although it hints that his control of Swansea Vale Works and 'command of Burmese ore concentrates' were strong points in his favour. He certainly pressed his scheme on the Government with forceful persuasiveness but, as he is unlikely to have been a match for W. S. Robinson and William Hughes in this art, a possible conclusion is that the Australians did not wish to press the issue as they were not prepared to take the risk of a slump in demand and prices after the war. Further evidence may, of course, come to light to disprove this theory later.

As a significant footnote to this story it must be mentioned that there was a third contender for Government aid for a large-scale zinc smelting industry. The History of the Ministry of Munitions mentions that at the end of August 1916 'a deputation representing British zinc mine owners, attended at the Ministry to urge the erection of additional smelting works'. Their case was that, with sufficient support, the British mines were 'capable of considerable development' sufficient to provide enough ore for a smelter to be erected at 'a convenient central spot' to produce 10,000 tons of zinc a year at a capital cost estimated at about £200,000.

'No action was taken on these proposals at the moment' but, as a consolation, a clause was put in the agreement of 1917 for setting up Avonmouth Works requiring that this works must be prepared to accept British ores 'up to 10 per cent of its capacity'.

This was a very real setback in any possible recovery of the zinc mining industry in Britain and ensured the hostility of the owners towards the Avonmouth venture, which expressed itself vehemently when the whole subject came to be debated in Parliament after the war.

CHAPTER FOUR

1917-1923
False Dawn–How the National Smelting Company Began

LIEUT.-COMMANDER J. M. KENWORTHY (Hon. Member for Hull (Central)) (later 10th Baron Strabolgi):

Another question which was not answered last night dealt with a very interesting subject, namely, the capital sum of £500,000 which was put into some factory at Avonmouth for making zinc concentrates or turning spelter into whatever spelter is turned into. I am not an expert and I do not pretend to have any technical knowledge, but I do know that this sum of £500,000 should be accounted for. It was put into this factory at Avonmouth which is, I think, somewhere in Wales (HON MEMBERS: 'Oh, oh!'). In any case the whole contract appears to have been a Welsh contract from beginning to end. I gather it was made by Welshmen on both sides. I am not, however, bringing any charge in regard to that: my present charge is only one of reticence as to the truth. This £500,000 I am informed and believe, on the good authority of people in the trade, was put into this factory at some period. What has happened to that factory and to the £500,000? Who has got away with the 'boodle'? This matter has never been discussed and this is practically a new service and we are entitled to an answer.

(*Extract from a debate on the Supplementary Estimates which took place in the House of Commons on 22 February 1922.*)

The founders of Avonmouth Works were not those who 'got away with the boodle' or reaped the ultimate reward of enterprise. They faded out of the scene after 1923. References have appeared in the previous chapter to the discussions which went on in Whitehall and the City in the first two years of the 1914–18 War about ways and means of establishing a zinc smelting industry in Britain and how the first few years of Avonmouth Works and of the modern British zinc industry will be for ever associated with the name of Richard Tilden Smith.

Four or five lengthy legal documents recount the succession of events which initiated the setting up of what is now Avonmouth Works but, to give a clear picture of these events, the chronological order in which they were officially executed will be ignored here.

The basic relationship between the Government and the Company formed to start the zinc smelting scheme at Avonmouth appears in the agreement between the Minister of Munitions and The National Smelting Company of 11 May 1917. This agreement begins 'Whereas H.M. the King has acquired a piece of freehold land on the east side of St. Andrews Road near Avonmouth in the City and County of Bristol . . . and is now (i.e. as late as May 1917) erecting thereon a factory and works for the manufacture of munitions of war and sulphuric acid and Whereas the Company is proposing to acquire a piece of freehold land to the south of the said piece of land acquired by His Majesty and on the same side of the road . . . and is proposing to erect thereon a factory and works for the smelting of zinc ores and concentrates to be purchased from the Ministry of Munitions as hereinafter mentioned . . . .'

Munitions, sulphuric acid, zinc ores and concentrates, zinc—this was the basis on which the original Company of the present Imperial Smelting Group was founded.

The double compulsion on the Government to find some outlet for the Australian concentrates which it was under contract to purchase and also to meet the vital need of zinc for munitions has been outlined in the previous chapter. The sulphuric acid and munitions side of the enterprise were part of a different but scarcely less urgent scheme. Sulphuric acid, which is a compulsory by-product of zinc smelting in developed countries, has played an important and often unwelcome part in the history of the zinc industry throughout the past fifty years and a history of the background of this involvement is set out in Chapter 10.

The danger of a possible shortage in the supplies of sulphuric acid essential to the production of explosives was realized in March 1915 by Lord Moulton, the distinguished Lord of Appeal who became Director-General of Explosives Supplies at the outbreak of the war. He at once established a Government Advisory Committee, as a result of which the sulphuric acid and fertilizer manufacturers were organized into a centrally controlled industry with a view to meeting national requirements.

The prime need at that time was for highly concentrated sulphuric acid of 100 per cent strength and also for oleum (100 per cent sulphuric acid containing sulphur trioxide in solution) for use in the production of high explosives. These demands led to very far-reaching changes in the British sulphuric acid industry.

There were sizable imports of oleum from America in the 1914-18 War but these proved insufficient to meet growing requirements so that a policy of new construction had to be initiated. Large oleum plants were constructed in connection with Government explosives factories and private manufacturers were encouraged to extend their existing chamber and concentrating plants.

The new acid plant at Avonmouth was therefore built with this demand for high strength acid in view but the Departmental Committee on the post-war position of the Sulphuric Acid and Fertilizer Trades, which was set up in February 1918, had the courage to face up to the probable consequences of this policy after the war, i.e. if the country needed a zinc industry based mainly on Australian sulphide concentrates, it must accept as permanent this inevitable addition to the country's total output of sulphuric acid, however embarrassing it might become when peace returned. So it made this very practical suggestion which must have reflected some of the thinking which went into the siting of Avonmouth Works a year earlier—'The acid produced from these ores will amount to about one-fifth of the gross surplus production. The Committee is of the opinion that every effort should be made to link up new zinc roasting plant with existing and efficient acid plant so as to avoid the needless erection of further surplus plant. The establishment of the zinc industry on a sound economic basis makes it essential that the roasting plants should not be far removed from the smelting plants on account of the difficulty and cost of transport of the burnt ore. It is also essential that the works should be situated in the neighbourhood of deep-water ports. These conditions rule out many of the existing acid plants. Nevertheless a number of existing plants are well situated, and proposals to employ such plants either in connection with zinc works in their neighbourhood or in connection with new zinc works to be erected on adjoining sites should receive the fullest consideration.'

This explains the plan to establish the new smelting works with Government assistance alongside the new Government plant at Avonmouth for producing concentrated acid and munitions. Avonmouth is certainly a deep-water port but the reasons why Avonmouth was chosen as the site of the industrial complex are not precisely known, although it was probably because it was remote from the East Coast bombing raids.

The official history of the Ministry of Munitions says that the original idea was to site the new roasters and zinc smelting plant near the Government acid and munitions factory at Queens Ferry but that Tilden Smith persuaded them that Avonmouth would be a more suitable site. As events turned out, although there was talk of other new zinc plants in other areas, Avonmouth became the only site where a zinc and munition works were established in close proximity and interdependence. There is no evidence that the Government ever attempted to establish an acid plant or a munitions works in the Swansea or Seaton Carew areas, which were the only other major zinc smelting areas at that time, and the other Government acid and munitions works which were established during the 1914–18 War, notably at Oldbury, Queens Ferry, Gretna and Greenwich, used sulphur as the raw material of acid production just as Avonmouth was compelled to do initially.

The subsequent history of this munitions factory at Avonmouth was brief and sinister. After the first use of mustard gas by the Germans in July 1917, development of a process, principally by Castner Kellner & Company, subsequently led to the decision to use the Government factory, which had recently been erected at Avonmouth and was originally intended for picric acid production, for production of mustard gas.

The History of the Ministry of Munitions continues the story of this project:

The H.S. (mustard gas) plant finally planned for Avonmouth was of a 500 ton per week capacity, the estimated total ultimate expenditure on conversion and construction of plant being £200,000. Some of the plant for ethylene production was brought from other works, and in order to get the earliest possible production, work was started up with make-shift plant. Production began on 3 July, 1918, and on 12 September the final plant was started up, the daily production being 5 tons.

The greatest weekly output reached before the Armistice was 125 tons, and the total output by the end of November was 417 tons. The initial difficulties of operation proved very serious. Upon two occasions the entire factory was closed down, and upon one of these the entire technical staff was incapacitated, partly owing to climatic conditions, partly to defects in ventilation.

The Government acquired the land for the munitions works, as has been mentioned, in 1917. The contract for acquisition of the land, on which the National Smelting Company's smelting works was to be built, was signed on 11 May 1917 by Tilden Smith, in his own name, and Philip Napier Miles. It mentions purchase of 100 acres of the Kings Weston Estate, at that time owned by the Miles family, for £20,000 i.e. £200 an acre. Covenants with the vendor stipulated that the land to be purchased was to be used solely for the construction of factories 'or connected offices and dwellings' and the purchaser was to compensate any tenant of the Kings Weston Estate whose tenancy suffered any damage or impairment of the fertility of the soil through the fumes arising from the purchaser's operations. The purchaser was to install modern means for the prevention of smoke and fumes and also was given the option to take over the tenancy of any tenant of the surrounding land who sought compensation for fume damage. These provisions were destined to assume great importance later. Tilden Smith was also given the option of acquiring within twenty-one years a specified quarry in the area presumably for building stone, and also 640 acres more of the Kings Weston Estate land. When this Agreement was 'adopted' by the Company ten days later about fifty acres of the purchased land and the Option on the quarry and on the 640 acres were excluded from the assignment so that the rights remained with Tilden Smith. Only the remaining freehold of the northern half of the land (about 50 acres) passed to the Company.

The remaining 50 acres was held by Tilden Smith in trust for the Company and not conveyed to it until 8 March 1924. The option on the 640 acres of Kings Weston Estate land, regrettably, was allowed to lapse and the Company acquired its present hundreds of acres of land round Avonmouth Works much later and in totally different circumstances.

The vendors, the Miles family of Kings Weston, are mentioned in Burke's Peerage and on the inn sign of the main hotel at Avonmouth, 'The Miles Arms', but there is no evidence of their motives for disposing of large acreages of land in this period. A large proportion of the land was subject to complicated mortgage and settlement rights.

The agreement for the physical construction of the works was signed on 5 April 1917, by Tilden Smith, in anticipation of the incorporation of the Company, with Robert McAlpine & Sons, Contractors and Engineers. They were to construct 'with all possible despatch' all necessary buildings, smelting furnaces, sidings, railway and tramlines, plant and machinery with all sub-sidiary works and plant required to treat 50,000 tons of concentrates per annum 'in the best and most approved and up-to-date manner' according to directions to be given by the Company or by Mr. P. E. Marmion who was at that time managing the Swansea Vale Spelter Company Works.

In addition, McAlpines were to erect on the Ministry land to the north of the Company's land and at the expense of the Minister 'furnaces for the roasting of ores and concentrates and for the extraction of the sulphur there-from'. These roasters were to be erected to the specifications of the Company, and the Company also agreed to 'provide an adequate and competent staff who will, under the instructions, directions, and control, and as Agents of the Minister, operate the Roasters . . .'.

The acid works were being erected under a separate contract between the Ministry and the contractor but it was also agreed that, after the war, the Minister was to lease to the Company 'such portions of the Sulphuric Acid Works as may be required and suitable for operation in connection with the Smelting Works'. The terms of the proposed ninety-nine years lease were appended to the agreement. As the lease was never implemented these terms are of little historical importance but several clauses show very clearly the importance to the Government of maintaining the right to reoccupy the acid works more or less at will and of ensuring that, in such circumstances, sulphurous gases from roasted zinc concentrates would always be available. This right even extended to the insertion of a clause giving the Government an option to acquire the Company's zinc works if it found itself deprived of sulphur gas from the roasters. The provisions for disposal of the acid were very elaborate and included requiring the Company to 'use their best endeavours to promote and establish other chemical factories in which sulphuric acid is used' and to supply such factories with acid on fair trading terms if they could not dispose of it to the Government or elsewhere. This, and other similar provisions prohibiting the joining of price rings and providing for fair wages, indicate that Whitehall may have been almost as active in controlling industry in 1917 as in recent years.

There are no references in the legal documents to the technical details of the smelting process to be built but there are many references to the roasters, Green Ore Store, Calamine Crusher and 'all necessary dust catchers and appliances for the economical and efficient handling of the ores and concentrates in connection therewith'.

The clearest summary of what sort of works Avonmouth was intended to be appeared in November 1919 in *The Times* Trade Supplement in a lengthy feature headed 'British Spelter Industry—New Developments—National Company's Enterprise'. It looks as though the correspondent was given a copy of the Company's expansion plans for Avonmouth and misunderstood that they had not yet been carried out.

Anyone reading the article would be justified in thinking that a gigantic smelter built on a lavish scale was about to operate. The programme contemplated an eventual yearly output of 70,000 tons of zinc and 100,000 tons of concentrated sulphuric acid. After referring to the acquisition of a quarry and the setting up of a special brickworks, the massive buildings and other civil engineering projects, the article goes on to describe the plant itself:

There are 20 roasters, each having 18 compartments connected by flues which will carry the sulphurous gas to acid towers for conversion into sulphuric acid. The distillation plant consists of 24 furnaces, divided into six blocks containing four furnaces each. Each furnace contains 384 retorts, or a total of 9,216 retorts in all. The gas producer plant consists of 12 batteries of five producers each, two batteries being erected between each pair of smelting blocks. The potteries comprise all the departments necessary for the manufacture of retorts, and the pot drying house, which is the largest in the world, has storage room for about 50,000 retorts, and is built of stone and brick.

Most of this project was never carried out.

What actually happened was that while some of the foundations for four distillation blocks were prepared, only one, Block A, consisting of four furnaces, was built during this period, but did not produce zinc until 1929. Of the four furnaces intended for Block B, two only were built, as an emergency measure, during the 1939-45 War. The remaining blocks were never built, part of the site and foundations being used eventually for the vertical retort plant.

Inevitably, in view of the background of the Australian contract outlined in the previous chapter, provisions laying down what types of concentrates the Company was to smelt took up over half of the principal agreement for the setting up of Avonmouth Works:

'For a period of ten years from 1st January, 1918 the Minister will direct the supply to the Company of 25,000 tons per annum of Australian (Broken Hill) Zinc Concentrates by such consignments as may from time to time be arranged and the Company will accept and pay for such concentrates on the same terms and conditions as those upon which the Board of Trade has agreed to acquire a larger amount of such concentrates under an agreement made or about to be made between Zinc Producers Association Proprietary Limited and the Board of Trade.' Ample reference has already been made to this notorious Zinc Producers Agreement.

Shipping of the concentrates was to be arranged by the Company in British ships and there were two lesser restrictions on the use of raw materials. The reason for one of these has been mentioned in a previous chapter. 'With a view to encouraging the mining of zinc ores in the United Kingdom the Company will be prepared to treat such ores in its smelting works up to a total amount not exceeding 10 per cent of their capacity.' This clause never had a chance to operate and Members of Parliament who protested at the ruin of the British non-ferrous mining industry in the famous debate of 1922 did not even mention it. The second restriction included was that 'The Company shall not without the previous consent of the Minister treat in the smelting works ores derived from Countries outside the British Empire to an amount in any year exceeding 15 per cent of the capacity of the Smelting Works for the time being.' Presumably this was inserted in deference to the Imperial scheme.

To compensate for being compelled to accept a price for concentrates fixed by the Government, the Company were guaranteed a minimum price of £23 per ton of 'Good Ordinary Brand' zinc to operate from the coming into operation of the Avonmouth Smelting Works until 1 January 1929. This clause was extended to cover all zinc produced by the Swansea Vale Spelter Company Limited as long as National Smelting Company remained in a position to control the output of the Swansea Works. The zinc from both works had to be held at the disposal of the Minister in the first instance but if the Minister did not require it the Government bound itself to make up any deficit below £23 incurred through selling the zinc on the open market. It will be remembered that F. A. Govett had asked for and was refused a guaranteed minimum price of £23 a ton when he and his colleagues considered taking on the British end of the Imperial scheme in 1916.

This was the setting in which The National Smelting Company was incorporated as a Private Company on 12 April 1917 and its first Articles reveal still more of the atmosphere of the time. There was an all pervading fear that, unless great vigilance was maintained, the industry would, by direct or indirect means, revert to control by German interests after, or even during, the war.

By Article 163 no one was to become a Director unless he was a British subject. By Article 164(A) any Director who ceased to be a British subject was to vacate office. Article 162 placed similar restrictions on shareholders—'In the event of any shareholder ceasing to be a British subject or becoming subject to the influence of any foreign state, body Corporation etc., under foreign control he shall disclose the fact to the Directors and his shares become liable to forfeiture.' There were other elaborate provisions to this effect reinforcing the central maxim laid down in Article 29—'Until peace shall be concluded in the present war between the United Kingdom and the German Empire, the Directors shall not register transfers of any shares *inter vivos* without the consent in writing of the Lords Commissioners of H.M. Treasury.'

These provisions are interesting as the embargo on the transfer of the Company's shares into foreign hands and on the appointment of persons who were not British subjects as Directors was destined to survive two radical reframings of the Articles in 1923 and 1954 and to last right up to the most recent reframing of the Articles on 1 May 1964.

Although in theory the cash for the new venture was to be provided 50 per cent by the Government loan and 50 per cent by Tilden Smith's private enterprise, in practice the Government put up most of the money in return for a theoretical 'whip hand' over the future of the whole scheme. In practice, also, the Government seem to have made very little use of their powers of control and, although the impression left from the legal documents is that Tilden Smith was given very little financial or other incentive to make a success of the scheme, except for the bonanza of the concentrates contract, he seems to have had an easy time with his obligations under the loan agreement signed on 11 May 1917 between the Company and the Minister of Munitions. By this agreement, provided the Company itself was prepared to raise £500,000 in equity share capital, the Minister agreed to advance to the Company sums up to a total of £500,000 against payments duly certified by the Minister's auditors as due for work done under approved contracts for the construction of the works or any future works extensions. The sum advanced was to be secured by Debentures issued to the Minister yielding interest at $\frac{1}{2}$ per cent over the Bank Rate of the previous half-year and with the exorbitant minimum for those times of 5 per cent. The Debentures were to be secured by a Trust Deed creating a specific first charge on the buildings, land, and fixed plant and a floating first charge over the remainder of the assets. There were provisions for annual redemption by means of a Sinking Fund to which not less than a third of the net annual profits was to be contributed with a minimum allocation equal to ten shillings per ton of zinc sold during the year. In the event, although all the £500,000 had been advanced and all the Debentures duly issued to the Government by 1 September 1919, there never were any profits or any zinc production during the currency of this agreement and, although there was a provision allowing Debenture interest to be paid out of capital, it is shown as 'accrued' in the Balance Sheet but was never paid after 1918.

Tilden Smith appears to have had an easy time also with his other obligations under the agreement. 499,997 out of the 500,000 £1 shares were allotted to him one shilling paid and remained one shilling paid until the first half of 1922 when two calls of five shillings each were made in rapid succession to reduce indebtedness to the Bank. The three other shares were allotted fully paid as Directors' qualifying shares, one each to Tilden Smith, S. C. Magennis and F. C. Heley. S. C. Magennis is variously described in official documents as a Bank Manager, Chairman of General Petroleum Company Limited, and a Director of various other companies including Bode Rubber Estates, Burma Mines and the Sunderland Iron Ore Company. He remained on the Board when the British Metal Corporation, The Zinc Corporation and other interests displaced Tilden Smith in 1923 and later became a Director of Imperial Smelting, an appointment which he held until his death in May 1947.

Jack Tilden Smith, brother of Richard Tilden Smith, joined the Board on 18 June 1917 while still serving as an Officer, and P. E. Marmion of Swansea smelting fame joined the Board on 23 May 1918.

There were no other accessions to the Board during the Tilden Smith era and the accession of the last four Directors produced no financial benefit to the Company as all that had happened by January 1923 was that 1,500 of R. Tilden Smith's one shilling paid shares had been distributed among the others.

According to the Minute Book, the first Board Meeting was held at the Registered Office at 4 Copthall Avenue at 2.30 p.m. on 20 April 1917. Apart from Tilden Smith there was only one other Director present at the time but it is possible that, following common practice, the Minute Book during Tilden Smith's era is a record of agreements between the Directors, couched in the form of Board Meeting minutes, rather than a record of actual physical meetings. This supposition rests on the fact that the minutes until 1923 are extremely brief and formal and provide no evidence of any discussion or argument on any subject.

Meanwhile, what happened on the two parts of the Avonmouth site (i.e. the Munitions factory and the National Smelting Company area*) between the end of the war in November 1918 and December 1923, when the Australian and other interests assumed control, is far from clear. Obviously munitions

---

*It should be remembered that until 1923 Swansea Works, which was producing zinc regularly, was owned by the Swansea Vale Spelter Company which was entirely separate from National Smelting Company.

manufacture ceased and with it the demand for sulphuric acid but it seems equally obvious that construction on the National Smelting site went on for about twelve months after the Armistice, to judge from the amount of money spent.

The first Balance Sheet is for the period ending 30 April 1918 and shows that construction expenditure up to that date had amounted to £278,116. The accompanying Directors' Report, which was the only one produced in this epoch, says that 'Constructional work has made steady progress in spite of numerous difficulties'.

By 30 April 1919, a year later, the total of construction expenditure incurred 'including Government Section and Stocks on the Site' had risen to £605,448. Only £17,000 had been spent on plant and machinery so the presumption is that most of the money was spent on site works.

There is a cryptic note in the Minute Book on 1 May 1919 which reads:

After full consideration of the position as affecting the Company's Works at Avon-mouth it was unanimously resolved that constructional works be suspended for the present and Sir Robert McAlpine & Sons were instructed accordingly.

However, a letter was sent to the Minister of Munitions

to state that pending instructions to the contrary, work on the roasters would be proceeded with.

It is certain that the roasters were not completed until November 1924 and, until then, when operating, the Ministry acid plant used sulphur dioxide from static sulphur burners.

The Balance Sheet that should have been produced in 1920 is not traceable but the next one, for the year ended 30 April 1921, shows that general construction work must have been resumed for a time as the net amount incurred to date on construction work, after deduction of £70,736 paid by the Government on account of their 'Section', had been no less than £803,828 i.e. over £300,000 in excess of the Government loan. In addition £40,488 had been spent on a 'Superphosphate Plant'. Part of this excess appears as a loan of £249,000 from Lloyds Bank and the remainder as 'Bank Overdrafts' and 'Credit Balances'. The Lloyds Bank debt was paid off by the two five shilling calls on the Ordinary Shares made in April 1922, but the M.P. who, in the debate in the House of Commons in September 1922 referred to the unfinished and derelict smelting works as 'that ruin at Avonmouth' might well have extended his description to the financial position and future prospects of the National Smelting Company at that time.

In fact the view expressed in the House of Commons in the Autumn of 1922 suggests that, as soon as war requirements had become a thing of the past, the

British mining lobby got the upper hand in Parliament and persuaded the Government that there was no necessity for extending zinc smelting in Britain or for keeping the Australian concentrates contract in existence if a way out of it could be found.

An M.P. remarked that: 'I hope whatever trouble the Government may be in now, they will not try to bring the raw materials to this country to be treated here. The opinion I formed at the time was that the best chance of an economic solution was to do the work in Australia and to produce the zinc itself on the spot.' To this the President of the Board of Trade (Stanley Baldwin) replied— 'The Hon. Member for Limehouse (Sir W. Pearce) said that he hoped we had no intention of manufacturing. We have not. We have no intention of manufacturing.'

The National Smelting Company could, therefore, expect no further help from the Government who acknowledged officially in 1923, by releasing the Company from its Debenture, that the £500,000 that they had put into the venture had been a complete loss. But the enterprise, almost at the last hour, escaped from the drift into liquidation. The Imperial idea had spread to non-ferrous metals and, as it coincided exactly with the Broken Hill producers' desire to keep the British contract alive, the Company was destined before many months had passed to receive a powerful stimulus to revival from the raw material producers and other metal interests.

CHAPTER FIVE

1923 – Empire Metal Interests Move In

THE POST-WAR SLUMP IN THE ZINC PRICE—SMALL BRITISH
ZINC ENTERPRISES FOUNDER—THOSE ANXIOUS ABOUT
'THAT RUIN AT AVONMOUTH'—LLOYDS BANK AND TILDEN SMITH
HIS ZINC INTERESTS BAITED WITH HIS BURMA MINES RICHES
THE IMPERIAL SCHEME REVIVED BY BRITISH METAL CORPORATION
AND SIR CECIL BUDD—W. S. ROBINSON DOUBTS BUT THE BOARD
FALL FOR INVOLVEMENT—SIR ROBERT HORNE
BRASS TACKS TALKED WITH TILDEN SMITH
THE COMPLICATIONS OF THE FINAL DEAL—THE GOVERNMENT
THROWS ITS HAND IN—RAISING THE PURCHASE MONEY
AND FACING UP TO THE PURCHASE COMMITMENTS
LITTLE SPARE CASH BUT A BRILLIANT NEW BOARD

REFERENCE TO APPENDIX I of this book (p. 201) shows that wartime 'scarcity' prices for zinc disappeared at the end of 1920 and that the level of prices ruling from 1915 to 1920 was not reached again for another quarter of a century, i.e. until just after the Second World War. This in itself is a surprising fact but when it is remembered that inflation and the abandonment of the Gold Standard were gradually eroding monetary values in these years, although not at the rapid pace to which post-war Britain has grown accustomed, it becomes an astounding fact of key significance in the history of the British zinc industry between the wars.

Much will be written in the remainder of this book about the far-reaching effects of this slump in the zinc price on the industry as a whole.

As regards its effect in the early 'twenties, the U.S. Department of Commerce statistics quoted in an earlier chapter show a decline in yearly average zinc production in Britain from 45,587 tons in 1916–20 to 29,561 tons in 1921–25 with exceptionally poor years at less than 20,000 tons in 1921 and 1922. Undoubtedly the boost given by the war to the U.S.A. zinc industry, which stepped into markets formerly supplied by the German and Belgian industries, and the complete absence of any form of protection for the British industry until 1932 were the main reasons for this depressing situation and the fall in price. It had visible effects in the closure of several small companies during this period—Dillwyn, Vivians, Pascoe Grenfell, Villiers and Brands Pure Spelter—whose history has been outlined elsewhere in this book. The result was that over half the independent zinc smelting companies which had existed before the First World War, disappeared and the fate of the remainder after the slump of 1929–31 will be described later. The subsequent history of the zinc-making sideline of a few large companies, such as Chance & Hunt and Stewart & Lloyds, who produced zinc during the First World War, is not known but it is unlikely that they continued this activity after 1923.

The independent companies who had been producing zinc were all of them concerns with little capital compared with the Government sponsored National Smelting Company which, however, had not yet produced any zinc at all and had not even fulfilled its obligation to complete its producing plant at Avonmouth.

Several concerns had at this time a substantial interest, directly or indirectly, in the future of this Avonmouth project—the Government because of its £500,000 loan, the Zinc Producers Association because of their long-term contract to sell zinc concentrates to the British Government, those who had thought out the 'Imperial Scheme' for smelting Australian concentrates round the Empire, and Lloyds Bank, who are said to have lent Tilden Smith several millions to finance his various schemes.

However, in the post-war atmosphere of jettisoning wartime projects as quickly as possible, the only people who appear to have been really worried about 'that ruin at Avonmouth' were Lloyds Bank.

Their worries, however, had a broader basis than Avonmouth Works and comprised several of Tilden Smith's other interests. In fact, the real starting point from which began the dominant development in the British zinc industry during the past fifty years—the emergence of one Australian-dominated Company, Imperial Smelting Corporation, as the sole component of the British industry—was one of those interests consisting of remote mines in the depths of Burma.

These lead-silver-zinc mines in the northern Shan States, not far from the Chinese border, had been worked intermittently by the Chinese for centuries but were derelict when A. C. Martin of Rangoon visited the area in 1891 and took some samples. For the subsequent and enthralling history of the development of 'Burma Mines' until they came under the Managing Directorship of W. S. Robinson in 1924 the authors are indebted to Mr. C. T. Fry of Burma Corporation Limited. Unfortunately, to maintain the relevance of the main text of this book to the British zinc industry, it has been necessary to relegate this account and additional information given in W. S. Robinson's memoirs, to Appendix III. However, from this story emerges the fact that Tilden Smith had been a Director of Burma Mines Limited since 1908 and probably acquired most of his enormous holding of 4,100,000 shares from Herbert Hoover who sold out in 1918. Herbert Hoover, as has been stated earlier in this book, was at that time a mining engineer and had been Joint Managing Director of Zinc Corporation with F. A. Govett from its re-incorporation in 1911. In 1929 he became Republican President of the U.S.A.

According to Lord Chandos, who is the sole survivor of the three or four who negotiated the future of Tilden Smith's zinc interests in 1923, the real attraction that these interests held for W. S. Robinson and the outside world was the Burma Mines shares, which were considered undervalued at their current market price. Tilden Smith, however, made it a condition of this deal, the purpose of which was to raise money to reassure the Bank, that his controlling interests in The National Smelting Company and The Swansea Vale Spelter Company should be included with the Burma shares in a 'package deal'. The reason for this was that, with zinc at about £33 a ton and with no Government money left to complete Avonmouth Works, these smelting interests were considered unsaleable by themselves.

Certainly the lingering wartime idea of an Imperial zinc production scheme influenced the outcome of the negotiations in that the Burma shares and the zinc interests were awarded eventually to the Australian interests represented by W. S. Robinson rather than to Brandeis, Goldschmidt & Co. Ltd. (now a subsidiary of Mercury Securities) who in the atmosphere of the time were regarded as too German in tone to qualify. This preference, when the deal between Lloyds Bank and Brandeis, Goldschmidt was nearing completion, was almost entirely due to the fact that the Imperial zinc production idea had been given practical expression in 1918 by the formation of British Metal Corporation Limited. This was partly through the inspiration of a wartime Civil Servant, Sir Cecil Budd.

Sir Cecil Budd, who died in 1945, appears now as a link between the pre-1914 German-dominated zinc industry and the new Imperial industry which was consolidated after 1923.

He was, before the First World War, senior partner in the firm of Vivian, Younger & Bond, metal merchants, who, among other interests, were members with Henry R. Merton & Co. Ltd., Metallgesellschaft, and the German controlled Australian Metal Co., of a consortium for the purchase of lead/zinc ores from the Broken Hill field. During the war he moved to the Ministry of Munitions where he was Controller of Metals and also served on several Committees of that Ministry, the Board of Trade and the Ministry of Reconstruction. There is no evidence that he took part in the negotiations with F. A. Govett and William Hughes over the Australian zinc concentrates contract and the establishment of a zinc smelting industry in Britain, or, later, with Tilden Smith over Government assistance for the Avonmouth project. However, it seems unlikely that he was left out of these discussions, particularly as he was Chairman of the London Metal Exchange from 1902 to 1928. Certainly he was caught up in the fervour for an Imperial zinc production scheme which was also stirring the emotions of the Zinc Corporation Board in these years. The result was the incorporation at the end of the war of the British Metal Corporation Limited which came into existence on 1 November 1918, with Sir Cecil Budd as the first Managing Director. The declared objects of this Company were and are 'to support and sustain the general trade in the United Kingdom in non-ferrous metals and to develop and extend the mineral and metal production of the Empire (Commonwealth)'. More precisely, as Lord Chandos, formerly one of its first Managing Directors, describes it, it was formed at the request of the Board of Trade to replace the German dominance in the Empire in non-ferrous metals with Empire self sufficiency. During these years British Metal Corporation remained representative of a broad section of world zinc, lead, copper and tin interests and was, until 1929, free from outside domination, although the Rio Tinto Company and Central Mining & Investment both had large shareholdings in it and representatives on the Board. It was also, from the time of its incorporation, the Board of Trade's agent for disposing of the zinc concentrates acquired under the Australian contract although, in this matter, it became the agent of The National Smelting Company from early 1924.

There was, in fact, ample enough communication between the Government and the Australian interests through Sir Cecil Budd and the British Metal Corporation to ensure that the Australian interests were given a chance to salvage the Government's interest in Avonmouth Works when Lloyds Bank raised the question.

There was also close communication between the Government and Lloyds Bank through the mediation of Sir Robert Horne, one of its Directors, with whom the crucial part of these negotiations for the takeover of Tilden Smith's zinc interests took place.

Sir Robert Horne, a few months later, on 11 December 1923, was to become Chairman of National Smelting and a few months earlier, in October 1922 had relinquished office as Chancellor of the Exchequer on the break-up of Lloyd George's Coalition Government. Born in 1871, the son of a Scottish minister, he won First Class Honours in Philosophy at Glasgow University and became a lecturer in philosophy and later a K.C., and businessman before, in 1917, he landed in a top wartime administrative post in the Admiralty and from there went into politics. In 1918 he became Third Civil Lord of the Admiralty and Conservative M.P., until 1937, for Hillhead, Glasgow. In 1919 he became Minister of Labour, in 1920 President of the Board of Trade, and in 1921 succeeded Austin Chamberlain as Chancellor of the Exchequer until he resigned with Lloyd George in October 1922. He appears to have been a close friend of Stanley Baldwin who had him elected to the Board of the Baldwin steel business (now merged in the nationalized firm Richard Thomas & Baldwin). He was created a Viscount in 1937, and died, still a bachelor, on 3 September 1940, a year after the outbreak of the Second World War.

His long and important association with the Empire zinc industry began with his involvement in these negotiations of 1923. He was drawn into them because Lloyds Bank, encouraged particularly by another of its Directors, W. A. Paine (father of Arthur Paine later Chairman of Burma Mines) was anxious to keep this Imperial zinc enterprise going and Sir Robert Horne was very much of an 'Imperial' figure. It was the age of 'Land of Hope and Glory' and the Empire Exhibition at Wembley (1924–25).

To return, however, to an even more formidable figure in these negotiations, W. S. Robinson conveys the impression in papers left by him that, although principal 'Empire' negotiator, he accepted the offer to negotiate with Tilden Smith, conveyed through Lloyds Bank, only reluctantly. According to him the position was that, although Tilden Smith, as mentioned previously, was very far from being in financial straits generally, Lloyds Bank had lent heavily to him in connection with his 'side' interests in the Avonmouth and Swansea zinc ventures, Burma mines and Kent collieries (his main interest at this time, it seems, was the reconstruction of failing mining companies). Lloyds Bank were, therefore, most anxious to sponsor a public company to take over some of these side interests. But from a purchaser's point of view, the background to the situation at that time was the drop in the zinc price to uneconomic levels immediately after the war and the mounting economic crisis of the 'twenties in Britain which was to pursue its way through the Coal and General Strikes up to the depression and crisis of 1929–31.

Nevertheless, this prospect does not seem to have deterred the Empire scheme promoters whom W. S. Robinson was representing. In 1923, major strikes and depressions could not yet be foreseen by the optimistic, and British Metal Corporation's annual reports for the years 1920 to 1923 showed continued optimism and limited success in the face of admittedly disappointing trade conditions. Even the knowledgeable and usually cautious F. A. Govett allowed optimism to influence him as was borne out in his statement to Zinc Corporation shareholders on 18 June 1924, outlining the motives for taking an interest in The National Smelting Company:

The smelting and acid business (i.e. Avonmouth and the Swansea Vale Spelter Company) was carefully examined on behalf of Broken Hill by Mr. (later Sir) Colin Fraser, Managing Director of the Associated Smelters (now B.H.A.S.) and the Electrolytic Zinc Co. (now E.Z.I.). His report was highly favourable. With the price at which the Burma interest was acquired and the prospects of that Company in addition to the prospects of the smelting company itself, I make no doubt that, failing a total European debacle, this will be a most successful company. This is another of the combined steps among the group for the production of Empire spelter, the finished product, from Empire material.

The first part of the negotiations, therefore, came to an inevitable end and the Board Minutes of British Metal Corporation of 27 November 1923, record—

Only a week ago we learned that a definite and acceptable offer had been made on behalf of an American Group. Thanks to the intervention of Mr. Robinson a preferential opportunity has been given to this Corporation.

The initial proposal was for a straight sale of Avonmouth, Swansea and the Burma shares to which Tilden Smith agreed, verbally and reluctantly. The following day, according to Lord Chandos, he changed his mind through fear of losing the Burma shares too easily and negotiations had to begin again with the object of finding some sort of formula which would safeguard Lloyds Bank and yet guarantee Tilden Smith dividends from the enterprise in future years if, after all, it proved a success.

The agreement that was eventually hammered out by W. S. Robinson, Sir Robert Horne, Captain Oliver Lyttelton (later Lord Chandos) and Tilden Smith was, therefore, an extremely complicated one. The terms of the deal fell into three main parts. 'Certain properties and assets' of the Swansea Vale Spelter Company, together with four million fully paid shares of ten Rupees each in the Burma Corporation were to be transferred to National Smelting for consideration worth £2,235,000. This consideration was to consist of 650,000 8 per cent Cumulative Preference Shares of £1 each and 1,000,000 non-voting Deferred Shares of 1/– each (with peculiar and important rights) to be created as part of the increased capital of National Smelting, and £1,535,000 in cash. The formal agreement, dated 17 December 1923, was between Tilden Smith joined with his creations, Intercontinental Trust (1913), Share Guarantee Trust, and The Trustees Corporation (India) Ltd. as vendors and National Smelting as purchaser, but when these new shares were issued on 7 March 1924 all the Preference Shares and the majority (625,000) of the Deferred Shares were put in the name of Lloyds Bank City Office Nominees Ltd. The remaining minority (375,000) of Deferred Shares were issued at various dates to British Metal Corporation, Metal Securities Ltd., Zinc Corporation, the National Sulphuric Acid Association, Gouldings and individual Directors in amounts of no significance and only 25,157 were allotted to Tilden Smith by name. The agreement provided that these Preference and Deferred Shares should be allotted to 'Intercontinental Trust (1913) Limited or its nominees'. Evidence of why these particular nominees were chosen, why a nominee 'front' should have been necessary, or who actually owned the shares for which Lloyds Bank were obviously acting as trustees and thus the secret of who actually held the power to decide the future of the Company, was destroyed in the war with the archives of Lloyds Bank Nominees Ltd. who continued to be registered as the holder of the majority of the Deferred Shares until 1929. If Tilden Smith really had incurred a large overdraft on account of his side activities, a possible explanation is that Lloyds Bank were holding these shares as security for the overdraft with the obligation to release them back free of their equity to Tilden Smith as the overdraft was paid off or to sell them to his order to raise funds to reduce the overdraft. From the account of the motives for the formation of Imperial Smelting given in Chapter 7 it seems certain, however, that Tilden Smith or some other outside interest and not Lloyds Bank still had the beneficial control of these shares in 1929 as it appears extremely unlikely that Lloyds Bank would have used the rights attaching to the shares to force a liquidation of this growing Company, which was the threat that the Deferred Shares held poised over the management until 1929. Another possibility suggested by the late Lord Baillieu is that these shares were being held by the bank in trust for the mining companies in the event of Tilden Smith agreeing to sell all or part of them.

The second part of the transaction was the purchase by National Smelting, from the Disposal and Liquidation Commission (successor to the Ministry of Munitions) and the Secretary of State for War, of the Government sulphuric acid works and roasters at Avonmouth for a net sum of £115,000 only. The agreement is dated 31 December 1923 and is interesting in that it provides further evidence of what had been going on at the Munitions Works since their construction in the middle of the war. A significant clause reads—'So far as any portions of the factory land that have been contaminated by mustard-gas or other deleterious substances and all buildings and plant therein that have been so contaminated the Commission hereby covenants with the Company to complete the demolition and cleansing. . . .' The Third schedule lists, among other items, 'the first 2 Ethylene Retorts' as due for demolition and visions of dichlorodiethyl sulphide (mustard gas) spring to the scientific mind. Other wording suggests that the land itself may not have been acquired by the Ministry until as late as 5 April 1917 and, although there is no evidence in available legal documents of the exact area of this useful land which National Smelting thus acquired on its existing Northern boundary at the end of 1923, a pencil note in an obscure but seemingly accurate file of the 'thirties suggests that it was 243 acres.

The third part of the transaction was equally fortunate for the Company. Included with the acid works, roasters, and land as part of the consideration for the £115,000 net paid by National Smelting under this remarkable bargain was the grant of freedom from the £500,000 loan made by the Government to the Company in 1917 as a help towards the cost of erecting the zinc smelting works. Admittedly no interest had been paid on the resulting Debentures since 31 December 1918 and the Government may have decided to write this sum off the books as a bad debt but it is remarkable that they decided to do this and to throw in the acid works, roasters, and land as well in return for only £115,000.

David Thomson in Volume 9 of the Pelican *History of England* has described how committed the post-war Lloyd George Government was to a policy of 'decontrols' and a 'return to normality' as quickly as possible. Selling off of wartime factories and 'war surplus' goods was one aspect of this and a probable explanation for the quick sale of the Avonmouth Munitions Works at a bargain price.

It was also provided by this Sale Agreement that all other relevant terms of the 1917 agreements between the Ministry and the Company were to be regarded as cancelled and this included the provisions dealing with the supply of Australian concentrates by the Government to the Avonmouth roasters. A new and more comprehensive agreement on Australian concentrates was concluded between the Company and the Board of Trade on 14 December 1923, and the main terms of this agreement form the background to later events in the Company's history.

To summarize the position at the end of 1923, The National Smelting Company found itself with half-finished smelting works at Avonmouth free from any financial or other obligations to the Government, with its Avonmouth boundaries extended northwards to include the sulphuric acid works and concentrates roasters and a large area of land, with the Swansea Works of The Swansea Vale Spelter Co. Ltd., now under its direct control,* and with a supply of concentrates assured until 1930.

It had also acquired over a third of the issued capital of Burma Corporation Limited. This brought to W. S. Robinson and Percy Marmion, Managing Director of the Swansea Vale Company, a novel and fascinating interest in the East, but the acquisition by National Smelting of these shares in 1923 was later to assume an importance probably completely unsuspected when they were acquired. Burma zinc concentrates did not arrive at Avonmouth until 1933 and the treasures of the mines, so soon to be overrun by the Japanese in the 1939–45 War and by Burmese independence afterwards, proved as elusive as the subtle charm of Burma; but the gradual sale of these shares for hard cash was the main source of finance for Imperial Smelting's expansion in the unpromising financial climate of the 'thirties.

As the cost of the acquisition of this mixed collection of 'assets', National Smelting had issued 1,000,000 Deferred Shares of 1/– each and 650,000 8 per cent Preference Shares of £1 each and had also to find a total of £1,650,000 in cash. The last Balance Sheet of the Tilden Smith era showed a minimum of cash and realizable assets and urgent steps were, therefore, taken to raise money by three different methods. At the Extraordinary General Meeting of 4 December 1923, which created the Preference and Deferred Shares, the authorized

---

* As a technicality it should be emphasized at this point that National Smelting did not take over The Swansea Vale Spelter Company. It merely acquired most of its 'properties and assets' leaving a 'shell' company which went into voluntary liquidation in 1924. Before 1924 the sole connection between Avonmouth and Swansea Works was the fact that Tilden Smith and his Trust Companies had shareholding control of each of the two separate companies concerned.

E

Ordinary Share Capital was doubled from £500,000 to £1 million. The remaining nine shillings a share was called up on the original 500,000 Ordinary Shares, of which 498,498 were still in the hands of Tilden Smith, providing £225,000. The 500,000 new Ordinary Shares were issued fully paid, providing £500,000 and £1,500,000 was raised by the issue of 7 per cent 1st Mortgage Debenture Stock secured on the Avonmouth property and the Burma shares and underwritten by the Union Underwriting Agency Limited, a subsidiary of Lloyds Bank, who found themselves left with most of the stock on their hands. The excess over the total sum of £1,650,000 due under the agreements was, therefore, £575,000 which, after deduction of approximately £100,000 for expenses incidental to increase of share capital, left working capital of less than £½ million.

New Articles which had been adopted in July 1923, were amended by a resolution of 19 December 1923 to define the rights of the three classes of shareholder which now existed. The resulting picture is so complicated that it is easy to understand why the whole of this structure had to be swept away by the creation of a new Company, Imperial Smelting Corporation, hardly six years later, to permit financial expansion to take place at all.

These complications arose, of course, from the deal which had been made with Tilden Smith and it would be tedious to attempt to unravel them here. The essential fact was that, while voting control still remained with the Ordinary shareholders and would extend to the Preference shareholders only if their dividends were two years in arrears, first refusal of all new shares created must be offered in the first place to the non-voting Deferred shareholders. For the issue of any further loan stock the sanction in writing of the holders of both three-quarters of the Deferred and three-quarters of the Preference Shares was required. As the Preference and Deferred Shares were created as part of the purchase price of the new assets and as, until June 1924, 50 per cent of the Ordinary Shares were also held by Tilden Smith and his associates, the whole effect of the arrangement, in theory, would have been to keep control on the expansion of the business in the hands of the vendors of the Swansea Vale Works and Burma shares (i.e. interests controlled by Tilden Smith).

As things evolved, Tilden Smith had transferred nearly all his Ordinary Shares by the end of 1924 and the bulk of them passed into the hands of British Metal Corporation and of Lloyds Bank City Office Nominees, who also held, nominally, all the Preference Shares and 62 per cent of the Deferred Shares. The actual details of the beneficial ownership of the shares held by Lloyds are, of course, undetermined as has been mentioned earlier in this chapter but, except for the peculiar feature of the Deferred Shares, the situation rapidly became less of a mystery. The Annual Returns show that the Preference Shares were widely held by the public by 29 July 1926, and that the majority of the Ordinary Shares had passed from Lloyds Bank Nominees to the public by August 1927 although Tilden Smith's Intercontinental Trust (1913) held nearly 25 per cent of them right down to the takeover by the newly formed Imperial Smelting Corporation in 1929. The shares were granted an official Stock Exchange quotation in March 1928 and the 'Ordinaries' were quoted at a price of around 30/– until 1929.

The Company which had 'gone public' on 21 June 1923, was reconverted to a Private Company in September 1931, two years after it had become a subsidiary of Imperial Smelting.

Against this complicated financial background the new Board brought to its task of making a thriving Imperial business out of 'that ruin at Avonmouth' and the ancient Swansea Vale Works a galaxy of talent but possibly insufficient working capital. At the first Board Meeting of the new regime at Baldwin House, Great Trinity Lane, in the City on 11 December 1923, Heley tendered the resignations of Tilden Smith and himself from the Board. S. C. Magennis remained on as a Director to watch over Tilden Smith's interests and Sir Robert Horne became Chairman. Sir Cecil Budd and Captain Oliver Lyttelton (later Lord Chandos) were elected as representatives of British Metal Corporation and Clive Baillieu (later Lord Baillieu) represented W. S. Robinson and F. A. Govett of The Zinc Corporation, both of whom were absent. Sir John Roper Wright and Sir John Davies, Chairman and Managing Director respectively of Baldwins steel firm, were elected to the Board. The reason for this strong Baldwin representation is not now apparent although it has been stated that Sir Robert Horne brought in his Baldwin colleagues to supplement the rest of the Board's lack of knowledge of metal smelting and heavy industry. Certainly Sir John Davies took a very active part in the surveillance of operations on the spot at Avonmouth and Swansea until his death in August 1927, less than a year after Sir John Roper Wright's death, and relations between The Zinc Corporation and the Baldwin element appear not to have been entirely cordial, probably because W. S. Robinson was also taking a close personal interest in the works at that time and found Sir John Davies another man of strong personality.

Two months later, after the conclusion of contracts for the disposal of the Company's sulphuric acid, R. G. Perry, of Chance & Hunt, who was Director and later Chairman of the recently formed National Sulphuric Acid Association, and Lord Wargrave of W. & H. M. Goulding, the Irish acid and fertilizer manufacturers, joined the Board. The Goulding representation has continued to the present day and the present representative on the Board of Imperial Smelting's parent Company, The Rio Tinto-Zinc Corporation, is Sir Basil Goulding, Bart.

In October 1926, Clive Baillieu joined the Board and, in November 1926, John Govett, later destined to be Chairman, succeeded his father who died in that month.

This most distinguished Board, which thus emerged in final form at the end of 1926, ruled the Company and the British zinc industry practically unchanged until the outbreak of the Second World War in 1940. It also had great influence on the Australian zinc and lead industry as its most prominent members were also Directors of the leading Australian companies in this field. Indeed, the unanimity of purpose between the highly individualistic Australian mining companies, which enabled the revival of the British zinc industry to be brought about, was mainly the result of the influence and efforts of W. S. Robinson and the Hon. W. L. Baillieu (father of Clive Baillieu) whose drive and financial acumen had been largely responsible for the initial development of the Broken Hill mining complex.

The success of the small National Smelting Company of 1924 in surviving the troubles of the 'twenties and the Great Depression and in expanding in unfavourable financial conditions in the 'thirties to become a leader in the world zinc industry was due very largely to the continuity of management provided by the Board under the leadership of the level headed intellect of the Scotsman, Sir Robert Horne, and the dynamic genius of W. S. Robinson. A large measure of praise is also due to the numerous unsung heroes who supported them away in the Plants in hygiene conditions which would not be tolerated today and to the steady support, particularly when cash was very short, of Lloyds Bank and the great stockbroking house of Govett, Sons & Co. The elements vital to an industry's success are indeed many.

CHAPTER SIX

1924-1929
A Fresh Start

THE BRITISH ZINC INDUSTRY IN 1924—A SET PURPOSE SLOW
IN DEVELOPING IN NATIONAL SMELTING—MARMION, ROBSON
AND IVEY—PROCESS DEVELOPMENT SET GOING BY THE NEED
TO IMPROVE WORKING CONDITIONS—DEVELOPMENT OF
ROASTING PROCESSES—THE ECONOMIC CLIMATE,
MAJOR STRIKES AND LACK OF CAPITAL SLOW DOWN EXPANSION
FOUNDATION OF THE NATIONAL SMELTING RESEARCH
DEPARTMENT AND EARLY EFFORTS TO FIND BETTER PROCESSES
THE FIRST SUBSIDIARY COMPANY—MODEST GROWTH IN PROFITS
HOUSING AND WELFARE SCHEMES IN THE 'TWENTIES

ALTHOUGH, AS A result of the events described in the previous chapter, The National Smelting Company became, at the end of 1923, by far the largest zinc smelting company in Britain from the point of view of capital resources, it is unlikely that it was at that time producing more than a minor proportion of the total zinc output of the country.

Apart from the English Crown Works at Swansea, which was at that time declining towards ultimate closure in 1931 and Tindale Zinc Extraction Limited, whose small Waelz operation was in a similar plight in the wilds of Northumberland, there were flourishing companies operating at Seaton Carew and Bloxwich.

The Seaton Carew Works was owned and supplied with raw material by Sulphide Corporation Limited of Broken Hill, Australia, and had an average of 8·35 horizontal distillation furnaces operating throughout 1924. The New Delaville Spelter Company at Bloxwich also had distillation furnaces operating in 1924 and, although it was said by one who saw them that 'they looked as though they had been designed after dinner and run up in a fog' their 'designer', Shortman, who came from Wales and founded the business, did manage to make them pay.

Although definitions of the term 'distillation furnace' vary in size and number of retorts and are therefore an unsafe guide in the absence of tonnage figures, Swansea Vale Works, which had so recently joined The National Smelting Company, had only ten furnaces in operation in 1924 and produced only 13,467 tons of zinc in 1925. The Avonmouth furnaces of the Company were not yet in operation.

However, there could be no doubt that, provided its able and distinguished Board concentrated on making the most of the opportunities which the comparatively large capitalization and the Australian concentrates contract had set before them, The National Smelting Company would very soon outdistance the other companies remaining in the British industry.

Unfortunately, concentration on a set purpose could not always be a feature of these years.

W. S. Robinson had much experience of mines and the international metal market but little of the technical details of smelting and, inevitably, his first reaction after the conclusion of the Tilden Smith deal was a prolonged visit to Burma Mines. Unfortunately, he decided to retain there Marmion, the former smelting boss of Swansea Vale Works, at a time when Marmion's experience would have been far more usefully employed in starting up Avonmouth smelting. Marmion returned from his managerial position in Burma within two years but not before the subtle charm of Burma had bred a distaste for working anywhere else. His advice was valuable over the coming years but the real work on the smelting sites was done by two new employees—Ivey, Works Manager (Smelting) and Robson, Works Manager (Acid).

The other obstacles to the real but limited progress that was made in the years from 1924 to 1929 will be apparent from a simple recital of what was achieved. Inevitably, so soon after the chance lumping together of assets which had hitherto been isolated enterprises and the gathering together of Directors with such diverse and substantial outside interests, it would be a perversion of the truth to pretend that there was a single dynamic policy from the beginning. This did not develop until 1929 but, surprising though it may seem at the present time when profitability is assumed to be the spur and employee welfare is taken for granted, a policy of improving smelting developed in these five years which had its origins in a desire to improve working conditions on the roasting plant. This was fortunate for the later progress of the industry as the super-session of horizontal distillation furnaces by mechanical mass producing processes would have been impossible without prior improvement in the preparation and roasting of raw materials for smelting.

During the 1920's the roasting plant at Avonmouth which originally consisted of two Delplace roasters, operated to roast concentrates to provide gas to feed the modified ex-Ministry acid plant. These Delplace roasters normally consist of seven superimposed hearths, some $16\frac{1}{2}$ ft. long and 4 ft. 4 in. wide, spaced 7 ins. apart. Drop holes at either end and centre of alternate hearths allow the ore to pass from each hearth to the one below by means of raking and pushing by hand 'rabbles' worked by men through ports at both sides. When well ignited the sulphur in the ore provided most of the fuel, but some heat from a coal fire was needed for the lower hearths if the sulphur content was to be reduced to less than about 8 per cent. Each roaster could roast 11,500 tons of ore a year, to give 10,000 tons of roasted material.

One Delplace roaster was started in November 1924 and the second in 1925. Some calcine (or roasted ore) was sent to Swansea but most of it was sold to continental smelters, two ships, the *Cato* and the *Ino*, plying between Avonmouth and Belgium with this traffic. The sales were subsidized by the Government on the basis of the relevant clauses in the Australian concentrates contract.

Three Delplace roasters had been installed at Swansea Works many years before (1912). W. S. Robinson's view was that human beings should not be expected to operate these hand rabbled furnaces so it may well be asked why National Smelting decided to install hand-operated roasters at the new Avonmouth Works when mechanical ones were available and used elsewhere—

particularly in the U.S.A. The answer must be that they were guided by the experience of Swansea Vale, who had tried various types of mechanical roasters and found that, in addition to causing excessive dust carry-over in the gas, none were capable of eliminating sulphur down to the $1 \cdot 5$ per cent maximum necessary for the charge to the zinc distillation furnaces. This was probably because mechanical roasters were developed in the U.S.A. for zinc concentrates of quite low lead content. With concentrates high in lead sulphide, as used in Britain, the easily fusible lead compounds gave trouble in the hearths owing to slaggy accretions which could be controlled and removed more easily in hand operated roasters. When the fine flotation concentrates from Australia came on the scene, trouble with slaggy accretions decreased, due to the lower lead content, but dust carry-over from the fine concentrates became excessive and the fineness of the product led to low recoveries in the distillation furnaces and great discomfort for the men working on the plants.

Clearly something had to be done to roast these fine flotation concentrates in an economic and hygienic way, yielding a product low in sulphur and in a physical form suitable for distillation.

Fortunately, a solution to this roasting problem emerged from the activities of Gilbert Rigg in Australia from 1917 onwards in the application of blast roasting or sintering to zinc concentrates, and also in the evolution in America of the Dwight-Lloyd sintering machine for lead concentrates. Rigg, who was a dapper and precise scientist with a goatee beard and an unfortunate habit of infuriating W. S. Robinson, visited Avonmouth about 1926 to discuss plans for a sinter machine. In sintering, air is forced through a layer of material several square feet in area and a few inches thick, the material containing sufficient sulphur so that it burns when ignited and sinters to a friable mass—5 per cent of sulphur is sufficient as fuel. In the original sintering machine, developed by Dwight and Lloyd, the grates or 'pallets' are arranged as a long endless chain passing over toothed wheels at each end. Material is fed over the pallet at one end, whence it passes under an igniter box and then over suction boxes where it burns and sinters, gas being ducted away to the acid plant. Sinter is tipped off at the far end of the machine.

Here then was a solution to the problems of roasting flotation concentrates—desulphurize only down to about 8 per cent sulphur in mechanical roasters and do the final sulphur elimination on Dwight-Lloyd sinter machines and the resulting product would be a porous material suitable for distillation.* Experiments at

*The final development, which did not come for some years, was due to the vision of Stanley Robson, Works Director of National Smelting, and was covered by his patent application of October 1927. This was 5:1 or straight sintering which does away entirely with

Model of the Champion (or English) process by sections. *By courtesy of Mr. Emlyn Evans*

Model of a Swansea Vale horizontal distillation furnace. *By courtesy of Mr. Emlyn Evans*

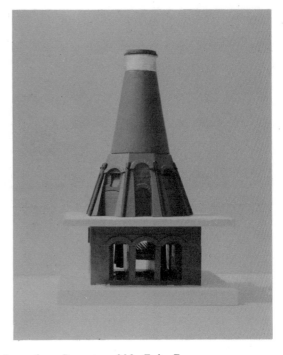

Model of the Champion (or English) process by sections. *By courtesy of Mr. Emlyn Evans*

Front end of a Dwight-Lloyd sintering machine (1926)

Firebox end—showing position of ore bed and platform

A view of Swansea Vale sintering plant (*c.* 1931)

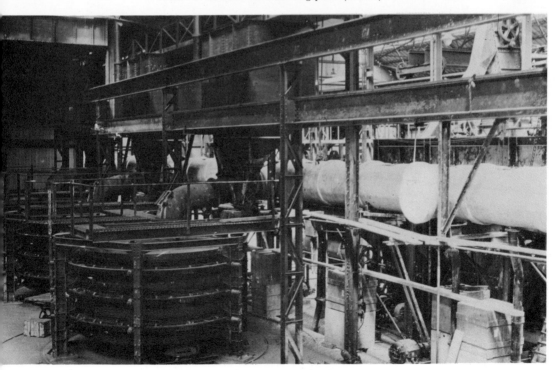

Barrier roaster furnaces, Avonmouth (*c.* 1926)

Model of a Swansea Vale horizontal distillation furnace. *By courtesy of Mr. Emlyn Evans*

Swansea using partially roasted Delplace material and fixed sinter pallets were successful, and plans were made at both works to install the Australian type Barrier roasters and Dwight-Lloyd sinter machines. In these Barrier roasters, developed from the de Spirlet roasters, but distinct from the de Spirlet in that no extraneous heat was used, alternate hearths only rotated, the rabbles being fixed to the underside of each hearth. This allowed very close inter-hearth spacing with a low drop from one hearth to the next which reduced dust carry-over and improved heat conservation.

Six of these Barrier roasters, which had been constructed by Sir William Arrol (Swansea) Limited, were commissioned at Avonmouth in 1926, the capacity of these and of the two Delplace roasters being given as a total 144 tons of concentrates per day. The same firm had also constructed six of these roasters at Swansea which were reported to be of 'a very high standard'.

In furtherance of the roasting scheme, the Board then approved, on 19 September 1925, a licence agreement with Huntingdon Heberlein under which a lump sum royalty payment of U.S. $11,000 per annum would be paid on installation of a Dwight-Lloyd sinter machine at Avonmouth.

Huntingdon and Heberlein, the founders of this firm, were two American metallurgists whose static process for improving the porosity of the charge for lead smelting preceded Dwight and Lloyd's idea of using a moving grate for the purpose.

However, although a small sintering machine was installed at Swansea in 1926 and at Avonmouth in 1927, it was not until 1929 that four large Dwight-Lloyd sintering machines were ordered and were installed, two at Avonmouth and two at Swansea, from 1930–32.

The main reasons for this delay seem obvious from the circumstances of the time.

In the first place, the economic climate was far from favourable to expansion. In 1925, the year before the General Strike, the mean average zinc price per ton was £36. 5s. od. but it had dropped to £24. 17s. 7d. by 1929, the major part of this fall taking place between 1926 and 1927. The Minute Book records on 21 October 1926 the decision that 'the Distillation furnace fires should be allowed to go out'. The Swansea furnaces only went out for a brief period until early in 1927. The two Avonmouth furnaces had not yet started production but were being heated up in April/May 1926 prior to start-up. They were

preroasting, all desulphurization being done on the Dwight-Lloyd sinter machines. As the concentrates start with 30 per cent sulphur and the optimum feed to the machines contains only 5 per cent, the new concentrate has to be diluted with five times its weight of sintered material —in fact the concentrate has to traverse the machine 6 times and, although this seems a long journey, the process has abundantly proved its worth.

allowed to cool down again as a result of the General Strike from 3–12 May 1926, and were not relit because of the prolonged Coal Strike of that year and generally adverse conditions. They eventually started up in January 1929.

The difficulty of obtaining additional capital for expansion was a further handicap. Construction work stopped during the Tilden Smith era and was resumed by the new management, some indication of the extent of this being given by the balance sheets.

Between the end of 1923 and the end of 1924 the value at cost of land, buildings, plant and machinery more than doubled to approximately £1½ million.

However, during 1925 'net additions' are shown as amounting only to £211,600. 1926, the year of the General Strike, saw less than £200,000 added, 1927 nil, and 1928 only some £15,000.

It is obvious that by the second half of 1925 the spare £½ million capital, which the Company had in hand at the end of 1923, was running out. On 15 October 1925 £25,000 was withdrawn from deposit and £30,000 of securities also realized to meet future liabilities. On 26 October, 250,000 Burma Corporation shares were sold and, in November, a further 750,000 and £20,000 more of the Company's other securities were designated for sale 'as and when further funds are necessary for the Company's programme of new equipment'. Sale of the whole holding of gilt-edged securities was approved on 17 December subject to similar conditions. On 18 March 1926 short-term borrowing was resorted to for the first time in the form of a £30,000 overdraft secured on the Company's holding of its own Debentures, and by 20 May this overdraft had increased to £60,000. On 16 June 1926 it was reported that a temporary loan of £40,000 had been obtained from an Australian source.

By 1927, however, the financial position appears to have eased as the withdrawal and cancellation of the 7 per cent Mortgage Debenture Stock began in June.

Yet another reason for the apparent slackening in construction work after 1926 is indicated in a minute of 18 February 1926:

The policy of the cessation of capital expenditure on any further extension of the distillation plant, including the proposed mechanical gas producers, pending investigation of other processes, was approved.

This was a statement full of significance for the future. Mention has been made in an earlier chapter of the inclusion by the Government of a clause in one of the agreements with Tilden Smith's company in 1917 stating that 'The Company will establish a department for research and enquiry into the best means of development of the manufacture of spelter, etc. etc.' This agreement had been cancelled by the later agreement of 1923 but the zest for research which has placed Imperial Smelting Corporation in its present pre-eminent position in the world of non-ferrous metallurgy and various branches of chemistry, has continued uninterrupted from its earliest beginnings. The Balance Sheets show that only small amounts of less than £1,000 a year were spent on research before 1924 but by 1927, when a formal Research Department was set up, these had risen to £12,518 and by 1928 to £14,133.

The years from 1924 to 1929 were conspicuous for the number of ideas on new processes and new sources of raw material supply that were discussed by the Board. The various mining options that were considered in such places as San Telmo (Spain), Iglesia (Mexico), and Tunis, came to nothing and the Australian concentrate contract remained the mainstay of the industry's operations but several of the new processes that were studied became established methods in the industry's scheme of production within the next decade.

This is not meant to be a technical history of zinc but it must be said here that the consciousness that the main established pyrometallurgical process of those days, the horizontal distillation process, must ultimately become scientifically and economically out-dated was a compelling factor in the Company's policy from 1924 onwards. The main available alternative in 1924 was the electrolytic process, which Amalgamated Zinc and the Zinc Producers Association had opted for when the Electrolytic Zinc Company was set up in Tasmania in the middle of the 1914–18 War. But the electrolytic process requires an abundant supply of cheap electricity, a condition which does not apply in the Britain of 1968 and applied still less in the 'twenties. Accordingly, various other processes were considered during this period for various stages of the zinc smelting process. The Lacell process involved conversion of zinc sulphide to molten zinc chloride and liquid sulphur, followed by electrolysis of the molten chloride to give zinc metal. This had several apparent advantages over the electrolytic zinc process which could have offset the high power cost in this country. The Remy-Roetzheim process was also considered. Both were tried on a pilot scale, but neither justified replacing the conventional horizontal retort process.

Attention was also given to the Waelz process in which material is reduced in large rotary kilns as a means of extracting the remaining zinc and lead values from retort residues or low value mineral in the form of rich oxides. Full scale plant trials were carried out in Germany in the autumn of 1926 under an agreement with Metallbank/Metallurgische Gesellschaft and Fried Krupp Grusenwerk, in whose works at Magdeburg a pilot unit was installed for the purpose. The process also seemed a possible alternative to roasting prior to distillation and plans were drawn up for its use in conjunction with the Remy-Roetzheim project which, as stated above, did not materialize. For this and other reasons, all thought of Waelz plants was then abandoned but the process was later to take its place, although perhaps not a very successful one, in the Company's history as will be described in Chapter 13.

Eventually, the vertical retort process of the New Jersey Zinc Company was destined to be adopted during the 'thirties but it did not make the major contribution to the Company's overall zinc output until after the Second World War and by 1955 the Company's own process—the Imperial Smelting process—had been developed to a stage at which it was capable of taking its place as the world's most modern commercial process for simultaneous zinc and lead production.

A significant reflection of the Company's forward looking policy after 1923 was the fact that its first subsidiary company, National Processes Limited, which was incorporated on 13 May 1927, had as its main object 'the development and exploitation of discoveries in relation to metallurgical and chemical practice'. Fifty-one per cent of the capital was allotted to National Smelting and 49 per cent to Stanley Robson*, who was at that time in charge of acid operations. He had been successful in adapting the Ministry contact plant from sulphuric acid production from brimstone to production from zinc roaster gases, and this and others of his inventions—in particular a vanadium catalyst for use in contact sulphuric acid plants—were the main stock in trade of this new Company.

As regards day-to-day commercial business, the years from 1924 to 1929 were not unsuccessful for The National Smelting Company in spite of adverse economic conditions. The net profit for 1924, which was the first year in which any profit was recorded, was £65,986, out of which the dividend on the 8 per cent Preference Shares was duly paid, which was the first dividend ever paid by the Company. In 1925 the net profit rose to £99,261 and 5 per cent was paid on the Ordinary Shares instead of the 8 per cent minimum stipulated in the Articles. Funds were insufficient for a dividend on the Deferred Shares which, under the Articles of that time, would be due to provide a dividend only after 10 per cent had been paid on the Preference and 20 per cent on the Ordinary Shares.

The profit figure for 1926 is impossible to disentangle from an overall figure given for 'Profit and Loss Appropriation Account' but appears to have been about £90,000 and, in addition to the Preference dividend, a dividend of

* Stanley Robson's 49 per cent share was bought out by National Smelting in 1937.

10 per cent was paid on the Ordinary Shares. The profit for 1927 was £176,092 and, in addition to 2 per cent extra on the 8 per cent Preference dividend, a dividend of 10 per cent on the Ordinary Shares was repeated. The profit for 1928 was £184,212 and similar dividends were repeated. The Profit and Loss Accounts, which are available for the years 1927 and 1928, show that 'Profit on Trading, Dividends, and Interest on Investments, after providing for Depreciation and Income Tax' for each of these years was in the region of £360,000.

These were fairly handsome profits for a small Company and undoubtedly sale of roasted concentrates was the main money spinner, apart from dividends on the large holding of Burma shares. At prices steadily dropping below £30 a ton there could not have been much profit from 15,000 tons or so annual production of zinc at Swansea Vale.

Disposal of sulphuric acid also brought little profit and was a source of anxiety from the earliest days. This has been a perennial feature of the industry's history and is considered in detail in Chapter 10.

Briefly, the country was faced with the problem of a serious excess of available capacity for acid production after the 1914–18 War. The Coal Strike in 1926 reduced outlets for acid still further but from 1927 demand appears to have recovered and consideration was being given to improvements to the acid plants, particularly by the adoption of the new refinements invented by Lurgi (Metallgesellschaft).

As has been emphasized already, technical improvements began in these early years of the reorganized National Smelting Company with the desire to improve the conditions in which men worked and it is appropriate that this chapter should end on a human note.

As early as 1917 the following clause appears in one of the Government's agreements with the Company regarding the proposed Avonmouth Smelter Works—'The Company shall make all reasonable provision for the health and welfare of its workpeople whilst employed on the Smelting Works'.

There is no record of how many people the Company employed directly during the Tilden Smith epoch. Presumably they were few, as National Smelting consisted then only of the unfinished smelting works at Avonmouth and the majority of those on the site, apart from those working in the Government munitions works, must have been building contractor's men. Swansea Works was still the responsibility of the Swansea Vale Spelter Company until 1923 and a passage from W. S. Robinson's memoirs indicates that amenities were almost non-existent there even in the late 'twenties. Largely as a result of his efforts messing and change rooms were improved at both works towards the end of this period, by which time (February 1929) the first available statistics show that there were 578 employees at Avonmouth and 560 at Swansea.

The field in which most action was taken, however, appears to have been housing. The first Directors' Report, issued in August 1918, contains this paragraph:

In view of the necessity of providing suitable housing accommodation for the labour to be employed in the Spelter and Acid Works when completed, the Ministry of Munitions are at present erecting 150 workmen's cottages at Shirehampton, in the vicinity of the Works, and a Public Utility Society has been formed for the purpose of laying out a new township on Garden City lines after the War.

There is no record of what happened to this Ministry project or to the workmen's cottages.

The Company's attention after 1923 was devoted first to the erection of twenty-seven dwellings for employees at Llansamlet (Swansea Vale). These houses were probably erected with the help of a Government subsidy in the region of £100 per house and were reported by 1 February 1926 as all leased to employees at fixed rentals. They later formed the nucleus of a 'Garden Village' run for a time by a subsidiary company of National Smelting.

Presumably because of delay in starting up smelting operations the 'general question of the provision of housing accommodation at Avonmouth' was not raised at Board level until May 1926. No specific proposals were made until May 1929, when W. S. Robinson proposed having seventy to eighty houses erected for employees on 10 acres of Bristol Corporation land by several alternative means and also the building of seven houses for foremen near the works at the cost of £450 each.

Meanwhile, on 1 April 1928, the Board had approved a Welfare Scheme which was to be based on an arrangement made with the Commercial Union Assurance Company and which would provide benefits, of generous proportions for those days, to all employees. The benefits were life assurance for each married worker and sickness benefit at £1 a week for any period up to one year, and payment of £1 a week in all cases of accidents. The total cost of this scheme was estimated at £2,000–£3,000 a year so the number of employees covered could not have been large. A Provident Fund scheme was also started at the same time.

Provision of life assurance and of sickness and accident benefit additional to State schemes was to be expanded into a regular system in later years but the response of employees to provision of Company housing was never very warm except during the housing shortage after the Second World War.

Finally, although the records are thin and the recollections of surviving employees of those days have grown uncertain, the beginnings of a 'Company spirit' are already discernible in this 1924–29 period, particularly in surviving legends told about various characters of the time.

From these it is obvious that there was already a spirit of devoted, if sometimes misguided, energy at work which was building the foundations, in this brief period of consolidation, for the much greater edifice which was to take shape with the incorporation of Imperial Smelting Corporation Limited in 1929.

CHAPTER SEVEN

1929–Why Imperial Smelting was Formed

THE DEAD HAND OF THE 1923 COMPROMISE—UNFORESEEN
IMPROVEMENT IN NATIONAL SMELTING'S PROSPECTS BY 1929
DANGERS IMPLICIT IN THE COMPANY'S VALUABLE SHARE PORTFOLIO
NEGOTIATIONS TO GIVE AUSTRALIAN MINING COMPANIES
WIDER OPPORTUNITIES IN THE BRITISH INDUSTRY
NATIONAL SMELTING BECOMES A SUBSIDIARY OF THE NEW
IMPERIAL SMELTING CORPORATION
THE WAY OPENED TO EXPANSION

THE BRITISH zinc smelting industry at the present day consists of one large Company only—Imperial Smelting Corporation. This was not the case at the time of this Company's formation in mid-1929 but became so within the next five years during which period Imperial Smelting acquired the assets of the remaining four zinc producers in Britain—English Crown Spelter Company, Tindale Zinc Extraction, New Delaville Spelter Company, and the Sulphide Corporation Works at Seaton Carew.

These acquisitions were not a difficult feat as, by 1930–33, the world economic crisis and the resulting slump in the zinc price had made life extremely difficult for these small companies: but events which took place within The National Smelting Company itself in 1929 were destined to have a much wider effect. In fact it is probable that the zinc smelting industry would not have survived at all in Britain if peculiar and local factors had not brought about a radical reorganization within the confines of the industry's largest Company at that time.

The reasons why the Board of National Smelting found it necessary to form another company, Imperial Smelting Corporation, to take over National Smelting as a prelude to expansion, rather than expanding the finance and activities of National Smelting itself, appear in Sir Robert Horne's speech to the shareholders of 31 July 1929. Fundamentally they stem from the restrictions imposed by the 'vendors' in the new Articles framed in 1923 on those who had been running the business since then. These restrictions have been outlined in Chapter 5. 1,000,000 Deferred Shares in National Smelting had been created as part of the price of purchasing the Swansea Vale assets and the Burma shares from some of Tilden Smith's other concerns. These shares, although they carried no voting rights at General Meetings, carried the sole right to first refusal of all new Ordinary Share Capital which might later be issued.

The Zinc Corporation, British Metal Corporation, Gouldings and the other Ordinary shareholders may have accepted this extraordinary arrangement so calmly in 1923 partly because an assumption commonly held at that time appears to have been that as the Australian zinc concentrates contract, made in 1917, had given rise to The National Smelting Company and Avonmouth Works, the whole enterprise would come to an end when the contract ran out in 1930. A passage in the British Metal Corporation Board Minutes of 7 November 1923, reads: 'So far as our enquiries have gone, it seems reasonably certain that in the event of complete liquidation after the completion of the Government contract in 1930, the capital would remain intact.'

It is clear, however, that by 1929 the situation had been transformed. As Sir Robert Horne said in that year, 'Looking back from this distance it is obvious

F

that the developments which have since occurred were never clearly foreseen—otherwise the existing Articles of Association and their restrictive conditions would not have been permitted to continue'.

By 1928 Swansea and Avonmouth, according to Sir Robert, were handling over 100,000 tons of zinc concentrates a year and producing about 30,000 tons of zinc and 100,000 tons of sulphuric acid and, although these three figures seem to be about 50 per cent in each case above the probable correct figures for 1928, prospects were undoubtedly brightening. The zinc price was still at a reasonably profitable level and, although an international cartel had had to be revived to limit production in response to threatening signs of a break in the price, no one appears to have had any inkling that an economic blizzard was about to begin for zinc and to last until the outbreak of another World War. Expansion was in the air and was encouraged by the fact that world output of zinc had risen from 1,130,000 tons in 1925 to 1,396,000 tons in 1928.

The prospects ahead of The National Smelting Company in 1928 were, therefore, vastly broader than the uncertain future facing the small smelting Company of 1923. The important point, though, in the minds of the Directors was that the potential sale value of the Company had grown even more than its prospects. This was not mainly due to the increased value of the smelting, acid and roasting plant at Avonmouth and Swansea Works and incipient optimism about the prospects of this young industry in the years between the General and Coal Strikes of 1925–26 and the Great Depression of 1930. The real value of the Company as a 'takeover' prospect in 1928 lay in its investments.

Again, to quote Sir Robert Horne—'In accordance with the terms of the Debenture Stock Trust Deed, the proceeds of such of our investments covered by it as were sold were paid to the Trustees and invested by them in British Government Stocks. The value of these stocks is now considerably in excess of the sum of £1,300,250 which now forms our Debenture debt. In addition, the market value of our remaining investments in the United Kingdom, Burma and Australia is not far short of £2 million.' The original investments of National Smelting consisted of the four million Burma Corporation shares which were shown in the Balance Sheet at the end of 1924 at £2,015,305 'at cost'. At the end of 1927 they were shown at a 'cost' value of £1,812,712 which indicates that the sales of shares to pay for plant expansion between 1924 and 1927 had amounted to about £200,000. At the end of 1928 the 'cost' value of all investments had risen again to £2,106,177 and, from Sir Robert Horne's statement, it is obvious that a hidden reserve consisting of some £1¼ million must have existed in the form of the difference between the cost and current market value of these investments. The earliest actual list of these investments that can be traced is dated 19 March 1931, and shows that, even after nearly two years of extensive expansion since July 1929, and the effects of the Wall Street crash of October 1929, the total book value of investments was still £885,333. Government securities accounted for £515,275 of this total and 2 million Burma shares (i.e. half the original holdings) for £293,000 of the balance. The rest were almost entirely holdings in various mining and metal companies.

This valuable portfolio of investments had a most important influence over the Board's policy from 1928 until the outbreak of the Second World War.

In the first place, when the metal interests on the Board were finally assured that the furnaces had started up at Avonmouth early in 1929 and were producing zinc of a satisfactory standard, they turned their attention to expanding in other directions. Capital was required for this and it was originally contemplated that this capital would be raised by the issue of further Ordinary Shares. This is why, eventually, Imperial Smelting was floated with authorized share capital of £7 million, although only just under £4½ million was issued initially which was used almost entirely to cover the purchase of National Smelting and Orr's Zinc White. After that, as events turned out, Imperial Smelting Corporation in the remaining nineteen years of its independent existence never issued any further share capital at all, apart from a mere £112,500 worth of Ordinary Shares issued exceptionally as the price for taking over the Australian mining companies' shares in Improved Metallurgy Limited in 1938. The slump started with the Wall Street crash at the end of October 1929, only three months after the formation of the new Company, and it was several years before conditions again became favourable for the raising of new capital on the market. Instead, most of the expansion programme that took place in the early 'thirties had to be financed from the investments acquired in 1923 and so greatly augmented in value between 1924 and 1929. In this the Company undoubtedly owned a great deal of its prosperity to the presence of John Govett (son of F. A. Govett) of the well-known stockbroking firm of Govett, Sons & Co. on the Board from 1923.

These investments had, however, more immediate significance for the metal interests on the National Smelting Board in 1929. At that time to raise more capital for expansion by issuing more shares would immediately have put the whole future existence of the Company and of these valuable investments in jeopardy. Under the Articles, as has been stated, any new share capital must first be offered exclusively to the holders of the Deferred Shares. Admittedly, as the British Metal Corporation minutes of 12 December 1923 stated 'Our holding of Ordinary Shares with that of the Australian group gave us a majority' (i.e. of the Ordinary Shares at that date) and it might therefore be thought that the threat that the Deferred Shares posed to the continued existence of The National Smelting Company is being over-emphasized in this book. It is clear, however, from Sir Robert Horne's speech of 31 July 1929, that in spite of the apparent decline in Tilden Smith's fortunes in the late 'twenties, this was a very real threat and in fact an issue vital to the continuance of the enterprise— 'a very substantial block of the Company's Ordinary capital and a great majority of the Deferred Shares were held by one Group. Had there been an extensive issue of new capital in the form of Ordinary Shares and had the Deferred shareholders exercised their rights the complete control of the Company (i.e. through holding more than 50 per cent of the Ordinary Shares) would have passed to them with power to wind up the Company voluntarily if they so desired'. The consequences of this for British Metal Corporation, The National Sulphuric Acid Association, Gouldings and the other Ordinary shareholders would have been considerable loss as Ordinary Share Capital was, under the terms of the Articles, due for repayment at par—20/- a share—whereas National Smelting shares stood at over 30/- on the market at that time. The Preference Shares, then held widely by the public, were also at risk as they stood at 28/- against 20/- par. The residue, amounting to some £2,383,000 after payment of the Debentures and creditors, would have gone to the 1,000,000 Deferred shareholders i.e. about 47/- for each one shilling Deferred Share.

Obviously the temptation to wind up the Company on the terms which a proposed increase in Ordinary Share Capital would have presented might well have proved stronger than the distant advantages of industrial expansion in the minds of any group of financiers, particularly if, as has been suggested in the case of Tilden Smith, they were in debt to the Bank.

Accordingly, the metal interests, led by W. S. Robinson, Managing Director of National Smelting, were driven to negotiate to remove this potential threat to the further development of the Company. They had in mind also the further incentive that the Government concentrates contract would run out the following year and that wider opportunities must be opened up for Empire raw material suppliers to take shares in the Company if their goodwill and interest were to be maintained.

No written details survive of the negotiations with the vendors but there emerged what Sir Robert Horne described as 'voluntary arrangement effected by the leading shareholders under which the ownership of the great majority of the Deferred and Ordinary Shares passed to the control of a number who indicated that they were prepared to carry through the readjustment of the Company's affairs which the position appeared to demand. . . . It is right that I should state that in this arrangement the Burma Corporation, South Broken

Hill, North Broken Hill, Electrolytic Zinc, and Zinc Corporation—all, as you know, leading producers of raw material—as well as British Metal Corporation are prominent participants.'

The arrangement did not, as might have been expected, take the form of a remodelling of National Smelting's Articles of Association to remove the restrictive Articles but consisted of an agreement to form a completely new Company, Imperial Smelting Corporation Limited, which started with standard Articles, Ordinary and Preference Shares with the usual rights, no Deferred Shares carrying special rights, and no limitation on increasing Ordinary Share Capital in favour of the Ordinary shareholders. Naturally no details survive of the course of the negotiations that led to this agreement but inevitably the terms offered to the holders of the Deferred Shares of National Smelting had to be generous to dislodge them from their privileged position. 825 £1 Ordinary Shares in Imperial Smelting were offered in exchange for 1,000 1/– Deferred Shares in National Smelting. This put a value of 16/6 on each 1/– Deferred Share or £516,675 on the large block of 625,000 still held in the name of Lloyds Bank City Office Nominees. 140 £1 I.S.C. Preference Shares yielding 6½ per cent were offered for every 100 £1 N.S.C. Preference Shares yielding 8 per cent (and quoted on the market at 28/–) and 150 £1 I.S.C. Ordinary Shares together with £5 cash (i.e. 2/– extra a share) for every 100 £1 Ordinary Shares of National Smelting (then quoted on the market at 30/–). A conversion offer was made to the National Smelting 7 per cent Debenture Stock Holders as a result of which just under a further million Preference Shares were issued in exchange and a few months afterwards 150,000 more Preference Shares were issued as part of the purchase price of Orr's Zinc White.

Apart from another 100,000 Ordinary Shares issued for cash at par to provide funds to cover the expenses of floating Imperial Smelting Corporation, no spare cash at all came from these initial share issues. The original intention was otherwise. The shareholders were told that 'at an early date—we hope sometime between now and the end of the year' additional Ordinary Share Capital would be raised 'chiefly to meet the capital requirements for the extension of the works recently authorized'.

The reasons why these hopes were disappointed have already been explained.

Unfortunately, a history of the British zinc smelting industry would be incomplete without including the arid financial details of the deal described in this chapter but the overall significance seems obvious. Control over the future of a Company which, by 1929, included the major part of the capital invested in the British zinc industry, had passed from the hidden financiers, represented by the Deferred Share majority, to overseas non-ferrous mining companies who were interested in maintaining the Company as a smelter of their raw materials. They did not have an absolute majority vote over the other shareholders as the National Smelting Ordinary Shares, on conversion to Imperial Smelting Corporation shares, were fairly widespread among the public, but blocks of shares held by the mining interests were sufficiently large to enable them to exert the preponderating influence on Imperial Smelting's policy. Inevitably, with the dominating Managing Director an Australian and Broken Hill the main source of raw materials supply, the Australian influence predominated and an 'alliance with Australian interests' was mentioned more than once in National Smelting Board minutes before Imperial Smelting came into being. The setting up of a British zinc industry with the active help and support of The Zinc Corporation of Australia, which failed to come about when the Government turned to Tilden Smith instead of F. A. Govett in 1915–16, became a reality with the setting up of Imperial Smelting in 1929. It eventually reached its logical conclusion in the merger of the two in The Consolidated Zinc Corporation in 1949 but the climate created by the Empire scheme, the interests of British Metal Corporation, and the financial position of Zinc Corporation probably prevented contemplation of such a merger in 1929.

The Board that carried on the business from 1929 was identical with the old National Smelting Board but, now that they were freed from restriction, the change in the emphasis of their policy is noticeable. The next ten years saw greater expansion in the Company's activities than at any time before or since until the Avonmouth expansion scheme of 1965–68.

CHAPTER EIGHT

1929–1939 Expansion

THE GREAT DEPRESSION DISPERSES OPTIMISM—THE ZINC PRICE
AT AN ALL TIME LOW—BRITAIN GOES PROTECTIONIST
BUT FULL PROTECTION FOR ZINC IS THWARTED
HARD TIMES FOR SHAREHOLDERS—THE EXPANSION PLAN
IMPROVEMENTS IN OUTPUT AND EFFICIENCY
ACID, LITHOPONE, ZINC OXIDE—PLANS FOR A COPPER SMELTER
BOOMING DEMAND FOR SULPHURIC ACID ADDS NEW WORKS
AND PROJECTS—ACQUISITION OF REMAINING BRITISH ZINC
SMELTERS—RESULTING COMPLEXITY OF OPERATIONS
IMPERIAL SMELTING MOVES OUT INTO A RECOGNIZED PLACE
IN WORLD NON-FERROUS METALS—IMPORTANT TECHNICAL
CONTACT ESTABLISHED WITH THE NEW JERSEY ZINC COMPANY

THE KEY TO the history of the British zinc industry during the decade between the incorporation of Imperial Smelting Corporation in 1929 and the outbreak of the Second World War in 1939 appears in a few sentences from the speech of Sir Robert Horne to the shareholders on 19 November 1936:

The foregoing review of operations which are subsidiary to our chief business will serve to reveal to you the wide range which our activities now cover. We have indeed divagated into many spheres never contemplated when we started this organisation in 1923. They have all developed naturally from our main purpose; and if any share-holders were prone to criticise the Board upon what may look like a diffusion of effort the answer lies in truth that, but for the profits which some of them have yielded, we should not have been in a position to pay even our Preference dividends during these last years. For the disconcerting fact stares us in the face that, in spite of an efficiency which is second to none in the world, it has not paid us to produce zinc in this country.

From the standpoint of its present highly capitalized and entrenched position in the nation's economy, it may seem incredible that on two occasions in the inter-war period zinc smelting might well have failed to survive at all as a British industry except for a remote possibility of continuing on a backyard basis. The first occasion was in the early 'twenties when, as has already been described, Government money and Tilden Smith's interest in the Avonmouth project ran out and the slump from 1920 onwards was jeopardizing the continued existence of the small zinc companies which had survived the war.

The second occasion was in the years from 1929 to 1932 when the world economic crisis brought with it a disastrous fall in the price of zinc from £26. 10s. 6d. in the first half of 1929 just before Imperial Smelting was formed to an all time low of £10. 11s. 3d.* sterling price (£8. 4s. 10d. gold price) in April 1932. Seventy per cent of this fall took place in the first twelve months of Imperial Smelting's existence and began with the collapse of the American boom in October 1929.

Although it was not foreseen at the time, it seems probable that if Imperial Smelting had not been formed with greatly increased capital just before this rapid collapse began, the old National Smelting Company would have had to close down like the other zinc smelting companies in Britain. Certainly the old Company, like the defunct companies, could not have afforded to 'divagate', to use Sir Robert Horne's expression, and this diversification and the Burma shares were the main factors that kept the new Company solvent until the outbreak of the Second World War in 1939.

* The lowest recorded L.M.E. price was £9 13s. 9d. in 1931, but this was also a true 'gold' price—see footnote on p. 187 for the difference between 'sterling' and 'gold' prices at this time.

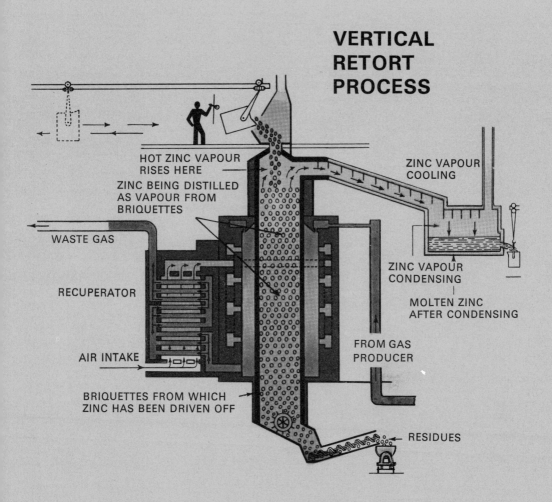

VERTICAL RETORT PROCESS

HOT ZINC VAPOUR
RISES HERE

ZINC BEING DISTILLED
AS VAPOUR FROM
BRIQUETTES

ZINC VAPOUR
COOLING

WASTE GAS

RECUPERATOR

ZINC VAPOUR
CONDENSING

MOLTEN ZINC
AFTER CONDENSING

AIR INTAKE

FROM GAS
PRODUCER

BRIQUETTES FROM WHICH
ZINC HAS BEEN DRIVEN OFF

RESIDUES

Vertical retort with pre-1954 condenser.

Charging a vertical retort

Casting molten zinc from a 2-ton ladle

VERTICAL RETORT PLANT

By 1932 British industry generally was starting to emerge from the slump and, in the hope of stimulating the process, Britain, and with it the British Empire, went protectionist, a movement culminating in the Ottawa Agreements of that year. As Neville Chamberlain, Chancellor of the Exchequer, said, the Government proposed 'by a moderate system of protection scientifically adjusted to the needs of industry and agriculture to transfer to our own factories and our own fields work which is now done elsewhere'. The zinc producing industry received the protection of a 10 per cent *ad valorem* import duty on foreign zinc entering the country but as a result of subsequent negotiations with Empire and foreign producers for a new international agreement the Import Duties Advisory Committee reduced this shortly afterwards to 12/6 a ton in spite of the Company's protests. The Committee admitted that the result of this compromise might be the opposite of Neville Chamberlain's intention— 'We recognize that the rate of duty proposed is low, and that in the absence of agreements with overseas producers to control production and sales, it may prove inadequate to safeguard the home industry'—and, after consoling itself with several sentences of Civil Service English expressing the pious hope that the Company's level of efficiency would suffice, nevertheless, to defeat the foreigner, left the door ajar for a renewed appeal if this hope were disappointed. The Company pushed hard at this door right up to 26 May 1939 when an increase in the duty from 12/6 to 30/– was granted but only just before the war suspended international dealings in zinc.*

These negotiations will be described in their proper context in Chapter 15. The overall result was that, after being hit like most other industries, by the great depression of 1929–32, the effect of inadequate protection exposed the zinc smelting industry to the full blast of the fluctuations of a depressed London Metal Exchange price right up to the Second World War and prevented it from earning anything more than a derisory return on its main product.

The tragedy was that there was, from 1932 until 1939, an increasing demand for zinc metal in Britain, particularly for galvanizing, motor cars, and armaments. Failure to obtain protection from imported zinc, arising from the excess production capacity of overseas producers, prevented the British industry from taking advantage of this growth in demand to build up capital and reserves as

---

*The reality of the situation is brought out in a draft press announcement, prepared by the Company in September 1938—'The Corporation's chief investment is in the zinc smelting industry which, owing to the pressure of competition from external producers in the British market, and the consequent heavy fall in the price of the metal, is at present unprofitable. In addition to the whole of the Seaton Carew and Delaville plants, it has been necessary to close one half of the horizontal retort distillation plant at Avonmouth and 40 per cent of the capacity of Swansea'—This was written only a year before the outbreak of the 1939–45 War!

The vertical retort plant at
Avonmouth and its product

a solid foundation for the future. Instead it had to live from hand to mouth financially in these years.

It is a great tribute, therefore, to the calibre of the leaders of the British and Australian zinc industries that, in spite of the economic blizzard that was to surround them both in Britain and Australia so soon after the formation of Imperial Smelting, they pressed on undaunted with their plans for expanding and diversifying the industry.

This could, of course, have been regarded as foolhardy. The shareholders of Imperial Smelting thought otherwise. They received a 5 per cent Dividend on their Ordinary Shares in 1930 at the end of the first year's operations and a similar Dividend at the end of the eighth year in 1937 (mainly owing to a speculative and unmaintained rise in the London Metal Exchange zinc price to over £37 in March of that year) but these were the only scraps of encouragement that were forthcoming in this decade. Several times there was the greatest difficulty in keeping up with payments of Preference dividends. Nevertheless, in spite of this disappointing return on their investment the records of Annual General Meetings are full of expressions of admiration for the Board's conduct of the Company's affairs. There can be no doubt that these were spontaneous and genuine, particularly as many of them were made as 'extras' after the speeches of those formally proposing and seconding votes of thanks. From 1929 to 1932, of course, shareholders in British industry generally had little to hope for after the Wall Street crash but, after 1932, the Company's plans for expansion and the manifest ability of its direction must have been sufficient to convince waverers that they had good hopes of eventual capital gain.

In the present age of forward planning the question therefore arises of what plans, if any, the Board had made or had in mind in 1929 for the future expansion of the British zinc industry.

A first general outline was given by Sir Robert Horne to the shareholders on 31 July 1929. As a preliminary he stated that the Government zinc concentrates contract had provided a source of cheap raw material in the decade while the industry was finding its feet and that the Company had made good use up to 1929 of this opportunity to establish the industry on a firm basis. This should enable it to stand on its own feet when the protection provided by the zinc concentrates contract came to an end on 30 June 1930. After that the Company would be in open competition with other zinc smelters for its raw materials. The explanation of this last sentence is that, as will be explained in detail in Chapter 15, the outlook for Australian zinc concentrates had changed vastly between 1914 and 1929. Instead of being a commodity for which new markets had to be invented, there were already more profitable outlets for these concentrates in 1929 than the British smelting works. In these competitive circumstances the more efficient smelter has a margin in hand over his costs of smelting to use in outbidding the less efficient smelter for purchase of concentrates and a measure of the abysmally low efficiencies from which Avonmouth and Swansea were rising in 1929 is contained in the following passage from the Chairman's speech made on 10 December 1931: 'Taking the year 1925 as a starting point—and putting the level of direct conversion costs of one ton of our chief product in our leading works at 100 in that year—we had by July of the present year (i.e. 1931) got our costs down to a figure of sixty. We . . . increased our metallurgical efficiencies by over 12 per cent in the same period.'

So large an increase would, by modern metallurgical standards, postulate an extremely low starting point.

The first objective in 1929, therefore, was to continue this improvement in efficiencies in order to make Avonmouth and Swansea a worthwhile outlet for Australian concentrates. With this went also the intention to increase the size of that outlet: 'We have authorized substantial additions to our plant both at Swansea Vale and Avonmouth and as new equipment is brought into work there will be a great increase in our operations and an appropriate return on our expenditure. The expenditure involved in our programme of new development falls but little short of £750,000 and provides for operations on a substantially larger scale than today in our zinc works.'

Beyond the immediate target of improving production efficiency and expanding outlets for zinc, both specific and more general objectives in other directions were mentioned in 1929: 'We hope in addition to expand in other directions—either by ourselves or in conjunction with others engaged in industries having similar or ancillary purposes. You will appreciate that we are directly interested in the production and use of acid. This covers a wide field and apart from transformation and use of zinc and its alloys we have a special interest in developing the production and use of lithopone and zinc oxide.'

Finally, after stating that the aim of the Company was to promote the development in Britain of the smelting of non-ferrous metals to the greatest extent shown to be economically possible, the Chairman added that, besides zinc, it was hoped to play a part in the production and treatment of lead and copper within the Empire.

Little was done about this third and final objective in the 'thirties and it need receive only passing mention here. Lead has played a minor and incidental part in Imperial Smelting's operations throughout its history but lead production in bulk remained established instead at Port Pirie on the Australian side of the Empire. As regards copper, on 9 September of the same year (1929) there was a reference at the Statutory General Meeting of the Company to preliminary steps taken for the erection in Britain of 'a modern and extensive copper smelting and refining works' but the project never came to anything. According to one of the Directors of the time it was defeated by failure to find suitable bismuth-free raw material.

It is obvious, therefore, from the records, part of which have already been quoted, that the long-term intentions on the zinc side of the business were fairly clearly defined in 1929.

By contrast it also seems most probable that, although the Board foresaw that other outlets, apart from direct sale, might be required to absorb the increasing quantities of the principal by-product, sulphuric acid, which would automatically become available with increasing zinc production, these acid outlets were not planned far in advance. There is no mention, for example, of the arrangement with British Aluminium to set up Aluminium Sulphate or with Fison, Packard & Prentice to set up National Fertilizers until the project is about to be put into execution. Secrecy of negotiations could, of course, be a reason for this but the most probable reason is that the demand for acid in Britain in the 'thirties fluctuated and, on the whole, far exceeded expectations held in 1929 whereas the market for Imperial Smelting zinc proved disappointingly unprofitable and limited for reasons already stated. Possibly because of the size of the acid using activities into which it entered, the position arose where inside and outside demand for acid could not be met in full from the contact plants at Avonmouth and Swansea which processed sulphurous gases from the roasting of zinc sulphide concentrates prior to smelting. As a consequence, small plants had to be brought in at the existing works and at works acquired at Newport and Pontardawe to meet the extra demand for sulphuric acid by producing it from brimstone and pyrites, purchased and imported specifically as the raw materials of sulphuric acid production. When the demand for acid declined, as it did, for example, early in 1938, the inevitable result was that some or all of this extra acid plant had to be closed down again. The sulphuric acid that the British zinc industry has contributed to the real wealth of Britain and the problems that Imperial Smelting has created for itself in disposing of this necessary by-product will be considered in Chapter 10.

This is one reason for the complexity of the history of Imperial Smelting in the decade after 1929. The first attempts to deal with the vagaries of the acid market added small works and plant to the list of the Company's assets. The second objective—the entry into lithopone and zinc oxide—added further works as will be described in Chapter 11. Still more were added as a result of the absorption by this one large zinc producer of some of the remaining small zinc works round the country which had survived the depression after the First World War. Part of this story has already been told. Apart from The Swansea Vale Spelter Company, whose disappearance as an independent entity has already been mentioned, English Crown Spelter in the same area was closed in 1930 and its assets were acquired shortly afterwards by Imperial Smelting for £50,000 and added to Swansea Works. Seaton Carew Works at West Hartlepool, which had been a tied outlet for Sulphide Corporation (Australia) concentrates, was acquired in October 1933. At the same time another small Company, The Delaville Spelter Company Ltd., was acquired in the Midland region with its works at Bloxwich. By this time, Imperial Smelting had become the sole zinc smelting concern in operation in Britain but at the cost of having zinc smelting operations divided between four separate and largely unrelated works at Swansea, Avonmouth, Seaton Carew and Bloxwich.

This complexity of Imperial Smelting's operations and operational sites was still further increased by the personalities inspiring its executive management. The two great Australians who produced so many of the ideas of these years, 'W.S.' and his son, 'L.B.' Robinson, were essentially Australian and adventurous in outlook. They had at their elbow, Stanley Robson, the Works Director, who had the outlook of the distinguished creative scientist that he was rather than that of the industrial manager. The ideas of these three added further ventures, some of them completely outside the field of zinc and its by-products.

Thus, for the separate reasons outlined in the preceding two pages, the operations of Imperial Smelting changed from zinc smelting with its associated sulphuric acid production at two sites in 1929 into a large complex of different activities carried on at one stage in the late 'thirties at no less than sixteen* sites. In addition, Imperial Smelting still retained an active interest in obtaining overseas zinc mining prospects quite apart from its Burma interest. The Second World War inevitably curtailed some of this activity but it was not until after the war that the attempt began to clear up some of the confusion and rationalize operations.

The 'thirties were, nevertheless, the period when Imperial Smelting's business ceased to be the 'skeleton in the cupboard' taken over with the valuable Burma Mines shares in 1923, and an inadequate outlet for the raw materials arising from the Government's wartime contract with the Australians. In spite of the appalling difficulties presented by the slump and inadequate protection,

*I.e. Avonmouth, Swansea Vale, Burry Port, Luton, Bloxwich, Seaton Carew, Widnes, Newport, Pontardawe, Haltwhistle (Tindale), Gasswater Mine, Cow Green Mine, Gowerton, Panteg, Port Talbot, Corby.

the enterprise moved out in this period from a purely local British setting into the international non-ferrous metals world as the expanding British zinc industry.

Undoubtedly, the frequent journeys of W. S. Robinson and other tycoons of the metals world between Britain and Australia by way of the U.S.A. were an important factor in this process. In particular, a most valuable contact was developed with The New Jersey Zinc Company (recently taken over by the Gulf & Western Company) which W. S. Robinson aptly described as an 'old established, wealthy, and extremely profitable concern'. New Jersey had, during the late 'twenties, performed the remarkable feat of bringing to the zinc producing industry three outstanding inventions, the exploitation of which by Imperial Smelting forms the main topic of the next chapter.

CHAPTER NINE

1929–1939
The Metal Basis

THE VARIETIES AND USES OF ZINC—BRITISH STANDARDS FOR ZINC
GRADES PRODUCED BY THE THREE MAIN PROCESSES AVAILABLE
IN THE 'THIRTIES—IMPERIAL SMELTING NEGOTIATES TO
LICENCE THE VERTICAL RETORT PROCESS OF NEW JERSEY
VERTICAL RETORTS AND REFLUXING PLANT INSTALLED AT
AVONMOUTH—IMPROVED METALLURGY LIMITED
THE VERTICAL RETORT PROCESS—LIMITED PROGRESS IN SMELTING
MORALE IN AN AGE OF MAMMOTH UNEMPLOYMENT
EXPANSION INTO OTHER PARTS OF BRITAIN—ACQUISITION OF
ZINC SMELTING WORKS AT BLOXWICH AND SEATON CAREW
MAKES IMPERIAL SMELTING SOLE ZINC METAL PRODUCER
IN BRITAIN—NEW JERSEY AND ZINC ALLOYS—ACQUISITION OF
ZINC ALLOY PATENTS AND MANUFACTURE AT MORRIS ASHBY'S
WORKS AND AVONMOUTH—TECHNICAL AND COMMERCIAL
OBSTACLES TO EXPANSION OF ZINC ALLOY DIE CASTING OVERCOME
WIDENING HORIZONS FOR ZINC DIE CASTING ALLOYS
PROFIT FROM THE CADMIUM IMPURITY IN ZINC CONCENTRATES
THE EXTRACTION AND USES OF CADMIUM

TO UNDERSTAND THE technical revolution in British zinc production which has been going on since the early 'thirties, a short section on zinc specifications may be found useful as an introduction.

All zinc used in industry has a content of at least 98·5 per cent Zn. There are, however, several main varieties of zinc and the variations all lie within the remaining $1\frac{1}{2}$ per cent. The uses of zinc dictate the purity required. These uses cover mainly galvanizing, brass, zinc oxide production, zinc sheet, zinc sprayed coating and die casting with purity increasing in this order, although the position of brass and zinc oxide on the list depend on the grade. Of more importance than the total zinc content is the level of individual impurity contents, mainly lead, iron and cadmium. For example, 'leaded' brass can tolerate a high lead content, while for 70/30 brass it must be very low. Carefully controlled impurity contents of all three are required for production of zinc sheet, both for 'rollability', stiffness, and electrochemical properties as applied to dry batteries. As cadmium oxide is dark brown it will mar the whiteness of zinc oxide, and lead must be very low in pharmaceutical grades. For die casting the permitted lead and cadmium contents are measured in parts per million; otherwise the castings will ultimately fail by inter-crystalline corrosion manifested by distortion and cracking. Hot dip galvanizing can tolerate fairly high impurity contents—in fact a certain amount of lead is preferred by some galvanizers.

British Standards Institution specifications for zinc have existed since the 'twenties. British Standards 220–222, which were first published in 1926 and revised in 1947, were not satisfactory as the impurity contents were unrealistic. Zinc of 99·99 per cent purity was not covered by a Standard (B.S.1003) until 1942. However, these standards were all rewritten in 1961 as British Standard 3436 which lists four grades of zinc designated Zn 1, Zn 2, Zn 3, and Zn 4 of minimum purities respectively 99·99 per cent (or 'Four Nine'), 99·95 per cent (electrolytic grade), 99·5 per cent (vertical retort quality such as Imperial Smelting's 'Severn') and 98·5 per cent. The first three grades usually command a price premium and, by selection within the four grades or by blending, all uses of zinc can be catered for.

The horizontal retort distillation furnaces were only capable of producing the grade of zinc known as Good Ordinary Brand (G.O.B.). In general this quality, which is the basis of the London Metal Exchange (L.M.E.) price, formerly ranged in impurity content, the world over, up to 2 per cent lead together with varying amounts of iron and cadmium, the former possibly up to 0·10 per cent and the latter in rare cases up to 0·20 per cent or over.

The general grade of metal produced by the process lacked uniformity and depended largely on the degree of purity of the zinc concentrates smelted (i.e.

the extent of the associated lead and cadmium contents) and also on the number of 'taps' or occasions on which the metal was drawn from the furnace condensers.

The characteristics of the concentrates have always been dependent upon the initial characteristics of the ore as mined and the efficiency of the ore beneficiation process in separating lead minerals from zinc minerals. The cadmium content is a fundamental characteristic of the ore and is not capable of separation by physical means.

Although Broken Hill zinc concentrates had become progressively better separated from the associated lead mineral galena throughout the 'twenties, the smelting activities of National Smelting during this period were based on concentrates stockpiled in Australia in the earlier war years, when separation processes were less efficient, and subsequently acquired under the Board of Trade contract outlined in previous chapters. These were high in lead and comparatively low in zinc.

Thus, in the horizontal smelting process the more volatile cadmium metal reported in higher concentrations in the early 'taps' while the less volatile lead reported in the later 'taps', leading to marked lack of uniformity in the metal produced as the distillation cycle of the furnace proceeded.

Several steps were taken by the industry to counteract these disadvantages. The various taps were segregated and made available to the consuming industries for which they were most suited. The heavily leaded later taps were refined by removing excess lead and iron by liquation in large settling furnaces holding some thirty tons of metal, compared with the very small liquating pots of earlier days. Thus, a uniform grade of around 1 per cent lead, 0·03 per cent iron and 0·04 per cent cadmium was made available, particularly for the general and sheet galvanizing industry. The other taps with lower lead and higher cadmium contents were suitable for the highly leaded brasses as some of the cadmium was later volatized during the alloying process.

However, apart from the technical disadvantages of the metal grade produced by horizontal distillation, the unpleasant and arduous manual labour necessary for the daily charging and discharging of furnaces at full heat of over 1,300°C was a problem of great concern to the industry's management as will be described later. Throughout the years many modifications and additions to the furnaces and auxiliaries were introduced to ameliorate these conditions, but basically the process was dependent upon intense human effort.

Meanwhile, a more scientifically sophisticated method of obtaining the metal from zinc concentrates had been achieved during the war years in the U.S.A. and Canada. About 1916, almost simultaneous work by the Anaconda

Company and the Consolidated Mining and Smelting Company at Trail, B.C., had lead to the successful development of an electrolytic process for the extraction of zinc. Shortly afterwards the Electrolytic Zinc Company of Australasia used this process for treating Broken Hill concentrates and their first commercial plant was in operation at Risdon, Tasmania, by 1918. The process had many advantages. Working conditions were far less arduous than with the traditional horizontal distillation process and it was capable of scientific control although many problems had still to be overcome. However, the factor of the greatest technical importance was that a uniform grade of zinc of 99·95 per cent purity was obtainable and that a later modification enabled zinc of still greater purity (99·99 + per cent) to be obtained.

The drawback of this process was that it required consumption of large quantities of cheap electrical power and thus was not suitable for adoption by the British zinc industry. This did not, however, prevent the higher grades of zinc that it produced from becoming popular generally. In fact, it is true to say that from the time of the development of the electrolytic process, the overall increase in the world's demand for zinc has been characterized by an increasing proportion of that demand being for the higher purity grades.

At the same time, therefore, as National Smelting was seeking to improve the operations of its horizontal distillation processes, there was an acute awareness of the increasing competition to be faced from the availability of these higher purity grades which could not be produced in its own smelters.

This impact had also been felt elsewhere.

The New Jersey Zinc Company of U.S.A. who operated a large smelter of horizontal retort furnaces at Palmerton, Pennsylvania, an area in the anthracite coal belt where no cheap hydroelectric power was available, had in 1929 developed, patented, and erected a new type of furnace which was continuous in operation, mechanically operated, and required a smaller labour force than horizontal distillation as it demanded far less arduous manual effort and skill. The problem of suitable manpower for horizontal retort distillation was, in fact, growing perhaps even more acute in the U.S.A. than in Britain, although the impact of this factor was similarly obscured by the world-wide depression of the period. This vertical retort process had some similarities in plant construction with vertical coke ovens. Batteries of eight or more retorts were erected in single blocks producing some 25/30 tons per day, i.e. about eight retorts producing 3 to 4 tons per day each, compared with 300 to 400 small clay retorts in one 'horizontal' furnace setting, each charged and discharged by hand and each producing 70/80 pounds per day. It had also been proved capable of producing metal of a markedly high grade of 99·7 per cent purity from the same Broken Hill mineral which produced a much lower grade in the horizontal furnaces. By the later development and adoption of a further process of high temperature fractional distillation in a 'refluxing' plant a grade of 99·99 + per cent—equal to that of the highest electrolytic quality—could be obtained as an integral part of this New Jersey vertical retort process. This grade of zinc is essential for the production of zinc pressure die casting alloys.

In fact, the adoption of the vertical retort process enabled Imperial Smelting to extend markedly the range of its zinc grades and also to make available a series of 'tailor-made' compositions for specific purposes, a facility which became of considerable significance during the 'fifties.

The written evidence of the time indicates that it was mainly foresight as regards the labour position and the feeling that a more economical process than the horizontal distillation system must be available, that stimulated the Directors of Imperial Smelting to investigate the vertical retort process almost as soon as it was patented.

The investigation was carried out by the Works Director, Stanley Robson, and some of his experts over an extended period in the U.S.A. in 1930–31, during which period trials with the raw materials likely to be available at Avonmouth took place. Simultaneously with these technical investigations, W. S. and L. B. Robinson initiated negotiations with the New Jersey Company as to the terms on which they might be willing to license their patents if the process was eventually accepted by the Imperial Smelting Board.

The process had so many technical advantages over the horizontal distillation method that there could have been little doubt that it would have been accepted on these grounds. However, the probable capital cost, by the uninflated standards of those days, was formidable. The whole plant, with its sixteen original retorts, together with necessary ancillary equipment, including the refluxers, and building, eventually cost over £400,000 to install as compared with a preliminary estimate of £230,000 in 1932 and the operating cost advantage over horizontal distillation as calculated in 1931 was so slight as to cause considerable doubts on the advisability of going ahead. Vertical retort operational costs were estimated in a report to the Board on 16 April 1931 at £7. 3s. 6d. per ton of metal as compared with £7. 8s. od. for horizontal distillation. W. S. Robinson's conclusion was—'in Mr. Marmion's opinion and my own, there is not sufficient justification for us to embark on the vertical retort. On his hopes, plus my figures, there is. I have requested Mr. Robson to fully reconsider the matter and let me have a supplementary report.' There is no trace of this supplementary report, but on 21 May 1931 the Board authorized W. S. Robinson to go ahead and complete negotiations with the New Jersey Zinc Company in

New York and on 15 June it is recorded that the Board approved the 'arrangement made in New York'.

The original intention had been to seek an exclusive option from New Jersey, for the whole of the British Empire, excluding Canada and Newfoundland— i.e. including Australia—and to build immediately 'one 50 tons a day' vertical retort unit at Avonmouth, the payment proposed being on a downpayment and royalty basis. This was in September 1929.

About a year later, when the effects of the slump and the fall in the zinc price were beginning to be felt, Hayes, President of New Jersey, was asked by W. S. Robinson whether the option could be reduced to one for the United Kingdom alone and based solely on a royalty assessed on tonnage and value of monthly production. Hayes was unwilling to reduce the rate of royalty but was willing to throw in for nothing a licence under the patents covering the newly installed New Jersey 'refluxer' process as an added incentive.

The royalty idea was subsequently discarded, a change of policy which delayed still further completion of the negotiations.

The final agreement, dated 15 December 1933, provided, briefly, that the United Kingdom patents covering the New Jersey vertical retort and refluxer process and first refusal of these rights offered in Australia, India and the Irish Free State, should be sold outright to Imperial Smelting for a downpayment of $340,000. Imperial Smelting was permitted to construct in Britain one vertical retort furnace plant of sixteen retorts with an annual production capacity of not less than 20,000 and not more than 22,500 tons. If before 1 July 1948, Imperial Smelting enlarged the capacity of the plant or built further refluxer capacity and put non-vertical retort zinc through the refluxer further instalments would become payable at an agreed rate per long ton of extra capacity required up to a maximum overall total in respect of the whole affair of $2 million. When this latter provision was put to the test by the 1938 proposals to raise the capacity of the Avonmouth vertical retort plant to 29,400 tons a year, New Jersey very generously agreed to adjust the amount of the instalment that thus became payable to allow for the fact that, over the past four years, the vertical retort plant had failed to produce the 22,500 tons a year expected by an average deficit of 4,041 tons a year. The only clause of this very friendly agreement which was planned to continue beyond 1 July 1948, was the one governing patenting and exchange of information on improvements to the process.

To carry out the vertical retort project a separate Company was incorporated on 12 November 1932 under the name of 'Improved Metallurgy Limited'. The reason for this was that outside financial participation had to be obtained, not so much because of the heavy cost of buying the patents and erecting the

Thus, compared with the horizontal retort process, in which recovery of waste heat and by-products is rudimentary, the vertical retort process collects and uses almost all the escaped zinc and waste heat.

The grade of metal produced by the vertical retort process has an analysis of minimum 99·7 per cent zinc, 0·2 per cent lead, 0·01 per cent iron and ·03–·06 per cent cadmium whereas the H.D. composition varied around 98·7 per cent zinc, 1·1 per cent lead, 0·04 per cent iron and 0·06 per cent cadmium depending on its position in the distillation cycle. After refluxing vertical retort metal became Crown Special—minimum 99·99 per cent zinc, 0·003 per cent lead, 0·003 per cent cadmium and 0·003 per cent iron.

There were thus three main grades now available to the Company and they were registered as 'AVONMOUTH' and 'SWANSEA VALE', 'SEVERN' (99·5 per cent minimum) and 'CROWN SPECIAL' (99·99 per cent). For the remaining five years of this pre-Second War period there was never any difficulty in producing Crown Special provided that sufficient Severn could first be obtained. In short, the refluxer worked well from the start and, with increasing demand for Crown Special as zinc alloy production passed from the research to the commercial stage in this period, refluxer capacity had to be increased twice. In 1936 a second refluxer column was added to the plant which brought its capacity up to 6,300 tons per annum. Subsequently, on 1 April 1938, a new plant of larger capacity (thirty tons a day) was also brought into operation. Total output capacity for high grade zinc had by then reached 12,000 tons a year as compared with the approximate capacity of 3,500 tons of the original plant. In 1936 Crown Special zinc was selling at a premium of 60/- to 70/- over 'G.O.B.' (Avonmouth and Swansea Vale), whereas it cost only 59/6 a ton to produce it on the refluxer. Apart from die casting, bright prospects were developing for this high purity metal in the field of zinc coating by spraying (using zinc wire), zinc rolling, brass, and electrolytic galvanizing.

By contrast, progress on the vertical retort plant was erratic in these years but, nevertheless, showed an upward trend. There appears to be no foundation whatsoever for the legend that later grew up among employees that this plant nearly ruined the Company in its earlier years. Admittedly it was bedevilled by retorts cracking, even at below the high optimum temperature required (1,350°C), and with two years' life only for each retort and nine weeks required for rebuilding a retort, it was a rare event at the best of times for as many as fourteen retorts to be in operation at the same time. Nevertheless output was running at a daily average of 48·6 tons during the last seven weeks of 1934, the start-up year; it exceeded 50 tons, its original target, in January 1935 and had reached 60 tons by April of that year.

By mid-1935 also, when the labour complement on the plant had reached 149, monthly sales of Severn zinc rose steadily from 2,253 tons in May to 2,634 tons in August but at a premium of £1 a ton only over the London Metal Exchange price for G.O.B. zinc.

No further substantial advance was made over these levels of efficiency during the remainder of this pre-war period and, after the first year's operation, cracking of retorts increased rather than diminished. This necessitated more frequent rebuilding or buttressing with a consequent drop in overall output but, by 1939, it was obvious that several retorts had been rebuilt on sounder principles to last a period of over two years.

A representative of the New Jersey Zinc Company who had been present at the start-up, returned in the second half of 1938 to discuss the reasons why the performance of the vertical retorts at Avonmouth had hitherto been 'below reasonable expectations' and also to see the new vertical retort developments at Oker in Germany. As a result of his visit the Board decided to cancel its previous decision to spend money on improving the horizontal retorts at Avonmouth and to spend it instead on improving output from the vertical retorts, mainly by increasing retort heights by 5 feet to 33 feet, installing a baghouse to collect some of the escaping zinc fume, and better charge preparation. The cost of the programme was estimated at about £60,000, and it was hoped to raise output by about 25 per cent to eighty tons a day. It was intended also to build a further battery of vertical retorts at Avonmouth to bring production up to 120 tons a day with a view to closing down horizontal distillation entirely on that site eventually. The 1939–45 War and the development of the Imperial Smelting process frustrated this plan to make Avonmouth entirely a vertical retort smelter. W. S. Robinson hoped that output would eventually reach six tons a day per retort (it did eventually in 1958) and recovery 94 per cent, but by 1939 it had reached only just over four tons a day.

None of the Company's zinc producing operations were more than marginally profitable in the 1930's and the new vertical retort plant was no exception. Separate accounts for the project are available for the first time for the six months ending 31 December 1938 and show a trading loss of £18,239 for that period. This, it can be assumed, was a smaller loss than that of the preceding three years but, considering the low level of zinc prices at the time, cannot be considered an abnormal loss for a new operation.

Meanwhile, Imperial Smelting continued to depend on the horizontal distillation furnaces for the bulk of its production and the history of these useful plants in these years is somewhat of an anti-climax.

After The Swansea Vale Spelter Company, with Government encouragement, had increased the number of its furnaces from four to twelve during the First World War, no further extension was ever made or ever appears to have been contemplated for this works. At Avonmouth, as has already been stated, two of the long awaited furnaces, first planned in 1917, came into operation early in 1929 and there was much talk at the time at Board and Annual General Meeting level of greatly extending smelting capacity there. Two further furnaces were ready to come into operation shortly afterwards but their start-up was delayed until January 1933, mainly because it was expected, to judge from various remarks in reports to the Board, that the negotiations to install the New Jersey vertical retort process would go through much quicker than they did. After this the price of zinc started falling rapidly which had the double effect of prolonging the vertical retort licensing negotiations, and also of reducing the incentive to expand zinc production. Hope for the zinc price was reviving in 1933 when the extra two horizontal furnaces were brought into operation but the behaviour of the price after 1932 and rising wages effectively discouraged any further idea of extending horizontal distillation. As W. S. Robinson wrote on 12 November 1932 —'We do not intend at this stage to proceed further with the completion of any further horizontal retort furnaces. We will adapt the buildings intended for the extension of the horizontal retort plant to a vertical retort installation.'

Production indeed continued by the horizontal process throughout this period but never at a level exceeding ten furnaces in operation at Swansea and four at Avonmouth. Market factors several times reduced Swansea to six furnaces and once closed the Avonmouth furnaces entirely. In normal periods, however, efficiency of output increased considerably during this decade, although efficiency of recovery of zinc from raw materials never varied greatly from 89 per cent. The figures for total annual output are deceptive as between 1933 and 1937 they never varied more than 1,000 tons either way from a mean of 40,000 tons and dropped to 35,000 tons in 1938. Clearly, as is borne out by the evidence of those who were involved in operating these furnaces at the time, the number of retorts actually producing zinc was varied frequently to increase or reduce output in response to market or cartel dictates. It is difficult to obtain figures to establish whether productivity per furnace was increasing during the 'thirties but figures worked out by Stanley Robson and submitted to the Board in September 1932 estimate that, taking 100 as the index of output per man furnace in 1922, output had increased to 116 by 1925, to 146 by 1928 and to 203 by the second half of 1932.

Improvements were introduced such as the three-tap system, larger retorts, prolongs, and the 'modernization' of six furnaces at Swansea beginning in 1938.

Indirectly also, improved working conditions on these furnaces arising from improved protection from heat during charging and tapping, the six day week, and better ventilation, must have contributed also to improved productivity.

As regards the vital factor of morale and its effect on production in the 'thirties, it is impossible to be dogmatic at this distance of time. The almost complete absence of strikes and walk-outs at the zinc smelting works was probably due mainly to a national unemployment figure which wavered around the level of three million and has left a sickening legacy of distrust of the 'bosses' as a class which may require several more decades of wise management for its elimination. In Imperial Smelting negotiations on a day-to-day basis between local managers and works committees chosen by Union members began in the 'twenties but formalization in written documents and official Councils did not set in until about 1936. The greatest product of these informal deliberations was an agreement giving the men twice the newly established piece-work rate for every ton by which they exceeded the previous output on time rates. This also was later codified and altered in detail on various occasions. Undoubtedly it stimulated production, but it would be unwise to venture an opinion on its effects on morale as this remains a matter of controversy among the survivors of that depressed pre-war epoch so long ago.

Geographical as well as technical expansion was also a feature of these years.

From the history of Imperial Smelting in the past thirty years it might be assumed that its deliberate policy was to centre itself in the South-West area of Britain. Except for pre-war arrangements over acid sales, this has never been the case. The entirely unrelated reasons for the siting of the two main works at Swansea and Avonmouth have been described in previous chapters and it is clear, from a letter of W. S. Robinson to the Chairman of the Board, dated 20 September 1933, that the desirability of extending zinc operations into the Central and Northern areas of the country had been under close consideration 'for some time past'. Very little time was wasted over carrying out this intention. It was reported, on 6 October 1933—only a fortnight after W. S. Robinson's letter—that Delaville Spelter Company Ltd. at Bloxwich in the Birmingham area and the spelter, zinc oxide, acid plants and other miscellaneous assets of the Sulphide Corporation of Australia situated at Seaton Carew near West Hartlepool had been acquired. No mention was made of these being the last zinc smelting concerns to survive in Britain independent of Imperial Smelting, but this was actually the case.

The Delaville Spelter Company Ltd. which was incorporated in 1901 had a nineteen acre freehold site 'directly on rail and with two canal connections'. It was estimated to have productive capacity for about 6,000 tons of zinc from secondary products, 1,500 tons of zinc dust and 2,500 tons of zinc oxide a year. The whole of the share capital of 19,685 £1 Ordinary Shares was acquired for 42/– a share, and the Chairman, Mr. Griffiths and a Director, Mr. Platten, agreed to stay on to form a Board with L. B. Robinson and F. A. Crew. This arrangement lasted until the end of 1944 when Griffiths and Platten retired and Delaville passed directly under the control of Imperial Smelting.

As events turned out, zinc smelting at Bloxwich played only an insignificant part in the Company's history during several brief periods of operation. Two furnaces are reported as operating in 1934 using zinc ashes as a raw material and producing 153 tons in September of that year, recovery of zinc from raw material being about 85 per cent. There was no significant change until the end of 1935 when improvements raised the recovery rate from 2·68 tons per furnace per day to 3·69 tons. Mixing in sinter with zinc ashes was tried the following year without success and in November 1936 the two zinc furnaces were closed down—for the same economic reasons, probably, that brought zinc smelting at Seaton Carew to an end three months earlier. Unlike Seaton Carew, however, the Bloxwich furnaces started up again for a brief and improved spell (over four tons a day per furnace) from April to December 1937. They were used again during the Second World War and subsequently scrapped.

The zinc oxide furnace at Bloxwich closed down finally, after intermittent operations, in August 1935 owing to the recent acquisition by Imperial Smelting of a 60 per cent interest in Fricker's Metal & Chemical Company under an agreement which debarred Imperial Smelting from producing zinc oxide elsewhere than at Fricker's two works.

The really worthwhile operation taken over with Delaville was the production of zinc dust.

Zinc dust is, as the name implies, zinc metal in the form of powder. It is made either by atomizing a stream of molten zinc with air or steam, or by condensing zinc vapour in a condenser designed so that droplets of zinc freeze before they have time to coalesce into a pool. Its principal use before the war was in the dyestuffs industry when zinc hydrosulphite, readily produced by reacting sulphur dioxide with zinc dust in water, was used to produce sodium hydrosulphite ('hydros') a powerful chemical reducing agent. Another use was for 'sherardizing', a process used for zinc coating small articles such as nuts and bolts which are not amenable to galvanizing because of dimensional change. Although used in the U.S.A. in zinc-rich paints, it was not until much later that zinc dust was adopted in this country for primers in which the dried coating consists largely of metallic zinc. This called for a dust of very small and closely controlled particle size.

An earnest attempt was made to build up the zinc dust business in the 'thirties in a market hitherto largely satisfied by imports. In September 1934 the plant is reported as closed down owing to accumulation of stocks. In October it was restarted on three furnaces. In April 1935 there is mention of a scheme for increasing the number to six and by March 1937 there were nine zinc dust furnaces operating. Eighty-six tons were produced in that month and there is mention later in the year of an increase in the selling margin from £8 to £9. 10s. 0d. a ton, although progress was not consistent. Market conditions enforced a long shut-down in 1938 but the country was able to put this plant to a very useful purpose on the outbreak of war and commercial progress was resumed in 1945.

The other works taken over in 1933 as part of the plan to spread zinc smelting activities into the Midlands and North was at Seaton Carew. From the zinc smelting point of view this project was also a failure and, unlike the Delaville Spelter Company with its zinc dust 'sideline', it brought no profitable by-product with it other than an additional quantity of sulphuric acid.

The origin of this works was that Sulphide Corporation Ltd. (of Australia) which is now entirely an Australian based producer of zinc, acid and fertilizers near Newcastle in New South Wales, owned, in the early years of this century, one of the mining leases at Broken Hill. In an endeavour to find an outlet for its zinc sulphide ores in England, it set up a Company under the name of The Central Zinc Company Ltd. and purchased a fifty-two acre site on the north bank of the River Tees near Seaton Carew. The project was, in fact, one of the pre-1914–18 War attempts to smelt the newly found sulphide ores outside the Continent of Europe. The assistance of a German metallurgist and German labour was used in the early years and up to 1930 Seaton Carew was a flourishing works, producing zinc known as 'Tees' brand, sulphuric acid (through a separate acid Company—The Central Acid Company until 1916), zinc oxide and (for a short time in 1927–28) hydrochloric acid.

The fall in the zinc price in 1930 put its zinc plants out of commission but, when Imperial Smelting acquired it through The National Smelting Company in 1933, the zinc oxide plant, two roasting plants, and two lead chamber acid plants were still in operation. The annual production capacity of the works at that time is given as 10,000 tons of zinc with capacity for roasting 25,000 tons of concentrates, 2,500 tons of zinc oxide, and 20,000 tons of sulphuric acid together with a hydrochloric acid plant. The works were reported to be well built and equipped in the main although smelting practice differed materially from that at Avonmouth and Swansea and production costs were higher. It was hoped that better standards could be introduced within two years at reasonable cost. The capital that had been expended on the works was estimated at £400,000 and the net purchase price, after excluding cost of stocks taken over and including the right to purchase the entire production of zinc concentrates from the Sulphide Corporation mine at Broken Hill for the rest of its life (it closed down in March 1940), was estimated at about £150,000.

The results of the first impact of Imperial Smelting on Seaton Carew were twofold. The zinc oxide plant was shut down in view, again, of the recent agreement between Imperial Smelting and Fricker's and plans were made to modify eight of the newer zinc distillation furnaces, built during the 1914–18 War, to take the size of retorts used at Avonmouth and Swansea. Removable condensers and Swansea Vale furnace practice were also introduced with the assistance of operatives sent up from Swansea but it is clear that Seaton Carew employees did not adapt themselves very readily to Swansea Vale methods. Four furnaces operated with monthly production rising slightly from about 330 tons and recovery of just over 80 per cent from July 1933 until August 1935. After that, inexorably, the low zinc price took its toll and the number of furnaces operating was reduced to two in August 1935 and to nil on 30 August 1936.

From W. S. Robinson's memorandum of 1 September 1936 it appears that in 1935 the intention had been to spend £100,000 on modernizing Seaton Carew Works if the Government were prepared to increase the inadequate protection given to zinc after the Ottawa Conference and to close the works if they did not. By 1936 it was clear that no increased protection would be forthcoming. Accordingly, the £100,000 scheme was dropped and the zinc furnaces were closed, but it was decided to continue the works in operation as an acid producer and to extend this side of it, partly to provide more employment. It was hoped that, at the worst, acid production would provide for depreciation and a small annual profit on the investment.

All that was left for the last third of the decade to 1939, therefore, were the roasting and acid plants. Concentrates were roasted on Delplace roasters, partly for Widnes Works (Canadian concentrates only) and partly to supply gas to the acid plants, until May 1937 when part of the plant was converted to the burning of brimstone. Seaton Carew continued merely as an acid producer until April 1941 when the furnaces were operated, again at a loss, until March 1945 to meet wartime needs; but these operations hardly justified the effort and money that the Company had spent on the smelting side of these works when it took them over. It had, however, been providing employment for over 240 men before closure of the furnaces in the 'thirties.

At the end of 1963 Seaton Carew Works was sold to Leathers Chemical Company.

Reference has been made at the end of Chapter 8 to the New Jersey Company's remarkable feat in the late 'twenties of bringing to the zinc producing industry three outstanding inventions within a short period of time. Imperial Smelting's adoption of the third of these inventions arose directly out of the ability to produce purer grades of zinc conferred on it by its prior decision to adopt New Jersey's vertical retort and refluxer processes.

During the early 'twenties in the U.S.A. a number of zinc alloys were produced and put on the market to serve the young die casting industry which was just beginning to expand. Zinc had a number of attractive features for the purpose. It had a relatively low melting point and was easy to melt and pour. It was very fluid, enabling intricate shapes to be cast. With the addition of aluminium, attack on steel dies and melting crucibles was reduced to a minimum, and strength and hardness could be gained by the addition of copper and other metals. As the advantages of die casting—this powerful new method of production engineering—began to be apparent the demand for zinc alloys increased to a considerable degree.

But although the alloys produced in the early days had many advantages they had one fatal weakness. In humid atmospheres corrosion spread rapidly through the metal along the grain boundaries until the structure became swollen and weak and could, in extreme cases, crumble into dust. It was little wonder, therefore, that zinc alloys were rapidly banned from the die casting field.

The New Jersey Zinc Company studied the problem in detail and found that the corrosive attack was closely associated with low melting point metallic impurities which were concentrated at the grain boundaries forming the alloy structure. If very pure zinc was used containing only a few parts per million of impurities, the corrosive attack could be reduced to negligible proportions. The refluxer which they had developed could produce zinc of the necessary purity and so the problem was solved.

Having solved the major problem they examined the various alloy systems in immense detail and eventually proposed six compositions (later cut down to two) which have formed the basis of all subsequent commercial development.

The essential component of all these 'Zamak' alloys is zinc of 99·99 per cent purity, the other three principal ingredients being aluminium, magnesium, and copper—hence the trade name 'Zamak', artistic licence providing K for copper. In spite of intensive efforts by Imperial Smelting and others no one has been able to improve, other than marginally, the original basic composition put forward by New Jersey in 1929 and 1930.

Imperial Smelting technologists went to the U.S.A. in 1931 to study this range of alloys and their applications, and their recommendation to confine promotion in Britain to two alloys only, No. 3 and No. 5, eventually became Company policy. There were obvious technical and commercial considerations which made it advisable to limit the number of grades used in a die casting plant.

Considerable resistance had to be overcome as memories persisted of the many failures encountered with the early alloys. A section of the Research Department was formed to specialize in Physical Metallurgy and a study began of the problems associated with the new alloys. Evidence of their stability, when made and cast to the necessary high standards of purity, was produced and the bad effects of contamination were also clearly shown. This evidence helped on a strenuous sales campaign, initiated by David Kirkwood (later Lord Kirkwood) and assisted by the preparation of a technical manual which was used as a 'Bible' by most British users of the alloys. A British Standard Specification was also prepared and issued which publicized the high standards of compounding and of casting which were essential for good service.

Use of these alloys has spread subsequently into numerous applications of castings both functional and decorative—motor car components, electrical and gas fitments, domestic appliances, vending machines, radio, television, toys, weighing, recording, and measuring instruments, and for precision engineering components during the Second World War.

However, legal as well as technical difficulties had to be overcome before this venture could be established on a sound basis in Britain. The negotiations for licensing the New Jersey patents on zinc base die casting alloys were not concluded until 1 March 1934, partly because of the pre-existing rights of Morris Ashby who were New Jersey's selling agents in Britain in this field. In earlier transactions Morris Ashby had acquired from New Jersey the right to manufacture zinc alloy using electrolytic grade zinc of 99·95 per cent purity as zinc of 99.99 per cent purity was not then available in Britain. The resulting alloy was therefore given the trade name 'Mazak' instead of 'Zamak'. The subsequent licensing by National Smelting on 15 December 1933 of the right to New Jersey's refluxer patents brought with it the assumption that 99·99 per cent zinc would be produced from the refluxer which was the grade specified in New Jersey's alloy patents. While, therefore, the alloy patents could be licensed by New Jersey direct to National Smelting just over two months later, further tripartite negotiations were necessary to secure the transfer of Morris Ashby's interim manufacturing rights to National Smelting. These were eventually secured in return for an agreement whereby Morris Ashby became selling agent of National Alloys, a 50/50 company formed by National Smelting and British Metal Corporation to develop the new enterprise.

Production of National Alloys' end-product therefore began and expanded in Morris Ashby's Works at Hackney. It was not until 10 February 1936 that National Alloys was able to make a reality of the transfer to it of the manufacturing rights by commissioning at Avonmouth a gas fired plant of some 5,000 tons annual capacity, using Crown Special zinc of 99·99 + per cent purity from the refluxing unit, and by closing down the plant at Hackney two months later.

By that time, following the issue of the first technical brochure in 1934, the major die casting companies had joined the list of customers and the volume of new die casting applications was steadily increasing, particularly among motor manufacturers. Sales were expected to exceed 3,000 tons in 1935 which upset previous calculations as to size of plant which should be erected at Avonmouth.

Accordingly the rated capacity of the new plant which started in February 1936 was designed at about 400 tons a month, but sales by July 1936 had already exceeded 490 tons. By January 1937 they had reached 660 tons and the records mention, naturally enough, that sales expansion was being limited by output.

The final development in this period was the introduction early in 1939 of a scheme for direct alloying whereby Mazak was made by direct admixture of aluminium-copper-magnesium into the molten zinc coming off the refluxer plant at Avonmouth.

Although sales rose as a result of vigorous promotional activities, only a few enthusiasts visualized the continuing rapid rate of growth which resulted in this and later decades.

Much has been written in the previous pages about the cadmium 'impurity' in zinc concentrates. The more profitable side of cadmium also merits attention.

All zinc concentrates contain some cadmium, those from Broken Hill, Australia, which have provided the bulk of the Company's raw materials for zinc smelting throughout its existence, having a cadmium content in the region of 0·2 per cent. Its existence in these concentrates was not suspected until the new electrolytic zinc plant erected by the Electrolytic Zinc Company at Risdon, Tasmania, at a substantial cost suddenly broke down only six weeks after its start-up in 1918. A few years later, soon after the adoption at Avonmouth of the Dwight-Lloyd sintering machine for desulphurizing of zinc concentrates, it was found also that cadmium was being concentrated in the material accumulating on the bars of the grates. These accretions were, and are, removed by rapping with chains—hence the term 'rappings'. Cadmium was also found to be present in the lead sludges recovered from the acid plant gas purification system and worthwhile concentrations of cadmium also appear in boiler oxide from the vertical retort plant, 'cadmium fume' from the refluxer, and residues from lithopone purification.

The problem of dealing with these materials was complicated. They still contained large amounts of zinc and lead which should be recovered but, in doing so, cadmium should not be allowed to build up to an excessive level in the zinc. Hence developed the urgent need to remove it from the smelting circuit.

There was a definite demand for cadmium, principally for yellow and red pigments, plating, and copper alloys, although its use in coatings for television picture tubes and in certain types of storage batteries was still to come.

Accordingly, soon after the installation at Avonmouth and Swansea of the new sinter machines, it was decided to build a plant at Avonmouth to process the cadmium 'rappings' and the cadmium arising from acid plant sludges. Cadmium at that time fetched only 2/– a lb. whereas in recent years it has risen as high as 24/– a lb. and is normally 10/– to 18/– a lb.; but it has always been an erratic business. It is an associated by-product of most zinc production operations and the market is always liable to be flooded with excess cadmium stocks when demand slackens with consequent rapid declines in price.

The cadmium plant at Avonmouth, which was largely a child of the Research Department, started up on an experimental scale in April 1936. Sales began in May 1936 and the purity of the product gradually rose to a guaranteed level of 99·95 per cent. Amounts produced fluctuated widely each month up to a peak of eighteen tons and sales were principally for export, although the home market had been protected by an import duty all this time which petitions by importers had failed to remove.

The process was a complicated one and there is no evidence now of whether or not the plant showed a profit or loss before the 1939–45 War. However, Imperial Smelting has had no alternative but to press on with production and watch the market carefully and it is perhaps better to regard the plant as a necessary step in maintaining zinc quality, rather than as a producer of cadmium, although this view has sometimes been questioned. Over the past

H

thirty years research has been active in keeping the process in step with changes on the zinc smelting side so that nearly all the cadmium from Avonmouth and Swansea is now channelled into one residual from which it is easily separated prior to conversion to metal.

An interesting development from this enterprise was the production until 1945 of cadmium-nickel bearing alloy by a process licensed from American Smelting & Refining and its sale through the Magnolia Anti-Friction Company to whom the process was licensed. The Bristol Aeroplane Company were much interested in this alloy at one time and it was also hoped that the motor car industry would take a bigger interest in it than they actually did.

CHAPTER TEN

1929-1939
The Sulphur
Complication and the
Foundations of
a Chemical Industry

ZINC SULPHIDE ORES INVOLVE THE BRITISH ZINC INDUSTRY
IN SULPHURIC ACID PRODUCTION—SULPHURIC ACID AS A
BAROMETER OF INDUSTRIAL PROSPERITY—HISTORY AND USES OF
SULPHURIC ACID IN BRITAIN—THE ALKALI ACTS OF 1906
TURN SWANSEA VALE WORKS TO ACID PRODUCTION
DEMAND FOR STRONG ACID AND OLEUM FOR EXPLOSIVES
MANUFACTURE IN THE 1914-18 WAR—CONSEQUENT SURPLUS
CAPACITY IN BRITAIN AFTER THE WAR—A GOVERNMENT
COMMITTEE RECOMMENDS SCRAPPING OF SURPLUS PLANT AND
RATIONALIZATION—FORMATION OF THE NATIONAL SULPHURIC
ACID ASSOCIATION—SULPHURIC ACID INVOLVES IMPERIAL
SMELTING IN THE CHEMICAL INDUSTRY—ACID DISPOSAL
CONTRACTS—THE SOUTH WALES AGREEMENT—THE SETTING UP OF
ALUMINIUM SULPHATE LIMITED WITH BRITISH ALUMINIUM
BEGINNINGS OF RESEARCH INTO FLUORINE CHEMICALS
ENTRY INTO FERTILIZER PRODUCTION—ACQUISITION OF BASIC
SLAG AND PHOSPHATE COMPANIES—AGREEMENT WITH FISON,
PACKARD & PRENTICE AND SETTING UP OF NATIONAL
FERTILIZERS LIMITED—NATIONAL FERTILIZERS LATER ABSORBED
IN FISONS—CUPRINOL, WOODWORM, DRY-ROT AND THE
SALES ADVENTURES OF AN ENTERPRISING COLONEL
LATER HISTORY OF CUPRINOL

THE SMELTING OF zinc sulphide ores on a large scale soon involved the British zinc industry, unwillingly, in the complicated politics of an old established industry. It is not intended to include a comprehensive history of the British sulphuric acid industry in this book as the authors would prefer to leave this task to those better qualified by knowledge and experience within the chemical industry.

A brief outline of the world into which the zinc industry introduced a new and cheaper type of sulphuric acid is, however, essential to the understanding of one of the biggest problems of the modern zinc industry.

It was Lord Beaconsfield who, appropriating an earlier dictum from Liebig's *Familiar Letters on Chemistry,* wrote that 'There is no better barometer to show the state of an industrial nation than the figure representing the consumption (i.e. in the sense of "use"!) of sulphuric acid per head of population.' There are no reliable figures available to show the production and use of sulphuric acid in the mid-Industrial Revolution era of Lord Beaconsfield but by the end of the 1939–45 World War production had reached nearly $1\frac{1}{2}$ million tons a year and has almost exactly doubled since then. This unattractive acid (H_2SO_4) which is in essence a compound of sulphur trioxide and water was, in the nineteenth century, used mainly in the production of fertilizers. J. B. Lawes in 1842 set up production of superphosphate fertilizers by treating natural calcium phosphate with sulphuric acid and by 1870 production of superphosphate in Britain had reached 40,000 tons a year. Other principal uses were for iron pickling and ammonium sulphate, hydrochloric acid, alkali and alum production, and in oil refining. The producing plants were almost entirely of the lead chamber type, i.e. plants in which sulphur, in the form of Spanish or Norwegian pyrites or brimstone, was burnt in air to give sulphur dioxide which, by reaction in lead chambers with oxides of nitrogen from nitre and with water, was converted into sulphuric acid of 65–75 per cent strength.

The impact of the discovery of the vast deposit of lead/zinc sulphide materials at Broken Hill, Australia, in the last quarter of the nineteenth century caused eventually almost as great a disturbance to the existing sulphuric acid industry of Britain as it did to its small zinc industry.

As has been emphasized, the modern zinc industry in Britain has been built up on the basis of abundant supplies of zinc sulphide concentrates from this Broken Hill field. On average just over 50 per cent of Broken Hill concentrates is zinc and the next main constituent is sulphur which, in recent years, has amounted to about 31–32 per cent. The sulphur content must be removed and the sulphide converted into oxide by roasting the concentrates before smelting can begin. The resulting sulphur dioxide must be disposed of. It has not been permissible since the first Alkali Acts of 1906 to discharge it into the atmosphere and the only alternative, therefore, is to turn it into sulphuric acid. In other words, the modern zinc industry must unavoidably draw on a cheap source of sulphur material not accessible to other sulphuric acid producers.

It is not surprising, therefore, to find that the possibility of developing a profitable outlet for the sulphurous gases from their smelting works was studied before 1914 by the Germans in control of the Swansea Vale Spelter Company.

As a result, in 1913 the neighbouring Briton Ferry Chemical & Manure Company agreed with the Swansea company to build and operate for a period of ten years a plant to turn the sulphur dioxide gases from roasting into sulphuric acid for their own use in fertilizer production. The plant was to be built adjacent to the roasting plant on the smelter works site and at the end of ten years ownership was to revert to the smelter. This agreement, together with the expansion of the smelter financed by the Government in the 1914–18 War, led to the building of chamber acid units on the Swansea Works site in 1914, 1915 and 1918 respectively. Total production at the end of the war from this works was approximately 50 tons (100 per cent equivalent) of acid a day.

The outbreak of the 1914–18 War revealed the urgent need not only for zinc but also for strong sulphuric acid and oleum. Oleum, a solution of sulphur trioxide in strong acid, is needed for the production of toluene which is essential for the manufacture of T.N.T. (trinitrotoluene) and of dyestuffs, in both of which German production capacity far out-stripped the meagre facilities available in Britain. The strong acid used for the manufacture of oleum can be manufactured either by concentrating weak chamber acid or by the 'contact' process which, in essence, uses a catalyst instead of nitre in the conversion process. The facilities available for production of oleum by either of these methods were meagre in 1914 as there had been no great demand for them. For this reason the first priority in the various new munition factories which the war brought into being was the construction of 'contact' plants to provide the necessary strong acid.

The new munitions factories at Avonmouth, Queens Ferry, Oldbury, Holton Heath, Gretna and Greenwich were, accordingly, all provided with 'contact' plants and, as has already been explained in Chapter 4, the Government financed the building of the roasters and acid plant at Avonmouth in the second half of the 1914–18 War with the set intention that the acid should go to their munitions plant and the roasted concentrates to Tilden Smith's zinc smelter as soon as it started up next door. By the time the war ended, however, only two of the six pairs of acid units which had been planned for the site to produce 20 tons of sulphur trioxide a day each (from brimstone initially) had been completed and only one had been put into use during the war. During the 'twenties Stanley Robson adapted this plant for use with roaster gases, which had been the original plan for it.

The urge to scrap all munitions plants after the first 'war to end all wars' was even stronger than after the second and, undoubtedly, the main reason why the Avonmouth acid plant escaped this fate was the 'post-war' provisions of the Agreement of 1917 with Tilden Smith's National Smelting Company. These provisions, as already explained in Chapter 3, owed their origin mainly to the long-term zinc concentrates supply contract with the Australian producers with which the Government had burdened itself in 1916. There was, therefore, no escape for the Government from the conclusion that this long-term contract also meant that there could be no return to the pre-war sulphuric acid situation. Room had to be found in the peace-time market for a quantity calculated by the Government Committee, set up in 1917, as likely to amount to about one-fifth of the gross surplus production of sulphuric acid after the war.

The Government Committee which reported in 1918 estimated that 'the national productive capacity before the war is difficult to assess but it was certainly larger than the actual output'. As regards consumption they held the assumption popular immediately after the 1914–18 War that the 'good old days' before 1914 would return rapidly, bringing with them use of sulphuric acid comparable to pre-war dimensions. They therefore estimated that the encouragement given by the Government during the war to every form of acid production was going to exaggerate the pre-war excess of capacity over demand to disastrous proportions. The expectation was that the excess would be equal to at least 35 per cent of pre-war output and that the consequences would be 'extreme dislocation in the acid trade of the country' and 'serious disturbance in the markets of the secondary manufactures which are produced by the majority of acid makers'. The Committee worried, in particular, that 'the production of sulphuric acid as a by-product in the roasting of concentrates is likely, unless carefully handled, to have serious consequences to established firms who will have to face competition from producers who will have to sell or otherwise dispose of their sulphuric acid irrespective of financial return owing to its unavoidable production'.

This was the heart of the problem.

The main remedy advocated by the Committee was the liberal use of Government compensation to encourage the scrapping of inefficient chamber plant, but the half finished acid and zinc works at Avonmouth was also viewed with disfavour—'We are of the opinion that the factory in question is not well situated for the sale of sulphuric acid as the peace-time consumption in the district is less than the probable output of the factory and is already fully met by existing works. The nearest large consuming areas both involve heavy carriage, which would probably make competition with local works impossible. It is therefore advisable that the acid should be utilized in manufactures carried out in the factory itself'.

One suggestion of the Committee that resulted in action was therefore the formation, with Government encouragement, of a 'National Sulphuric Acid Association'* on 7 April 1919 to rationalize the industry on 'a permanently prosperous basis' commercially and scientifically, having regard to the necessity of its maintenance as a key industry in relation to the needs of other industries particularly in time of war.

The section of the findings of the Government Committee which have been quoted here underline some of the difficulties which have faced the zinc industry over this aspect of its business since the end of the 1914–18 War. At any given moment since then, except during and for seven years after the 1939–45 War, there has probably always been excess production capacity available for sulphuric acid in Britain. However, it does not follow that the sulphuric acid price must have remained correspondingly depressed. The reliance of the vast majority of producers on raw materials (principally brimstone and pyrites) from varying overseas sources has been an influence on the sulphuric acid price far stronger than the normal play of supply and demand, and the price that the zinc industry receives for its sulphuric acid which is produced from zinc concentrates has to follow the market price for acid made from sulphur regardless of the factors affecting the price of zinc concentrates, e.g. a high sulphur price can to some extent compensate the zinc industry for a low zinc price. Another factor affecting the zinc industry alone is that it must go on producing acid automatically with zinc—even at times when it would be prudent for a sulphuric acid producer to close down owing to low prices.

The zinc industry has had, accordingly, to temper its policy on acid disposal to many contrary winds in the past fifty years and the most effective stabilizers in the storm have been found to be long-term contracts with customers and long-term projects for subsidiary manufactures based on the use of acid. The ability that a steady supply of sulphurous gas gives it to undertake long-term

*It is, incidentally, a measure of the extent to which Government attitude to private industry has changed in forty years that the joint sulphur purchasing activities of the National Sulphuric Acid Association were investigated under the Restrictive Practices Act in 1963, and again in 1966. The Association survived the hearings and has always had the unfailing support of Imperial Smelting.

contracts has given it the reputation of a steady source of supply while sub-sidiary manufactures have been the main reason for the Imperial Smelting's ever growing influence in the chemical field.

This entry into the chemical field has in itself brought problems vastly different from those encountered on the metals side of the business. In metals Imperial Smelting has been for nearly half a century the major and, since 1933, the sole producer of zinc in Britain and one of several big producers in the world as a whole. In chemicals it has been a small growing concern frequently competing against much larger British and overseas concerns formed primarily and not incidentally for production of chemicals in a world perpetually compli-cated by new processes, new patent rights and new selling methods. Most of these concerns have had their own sulphuric acid plants so have not provided markets for Imperial Smelting's acid. Some have used the low cost advantages that their large-scale plant conferred to keep Imperial Smelting out of their respective fields of interest. With others, in the days before the Restrictive Practices Act, it has been necessary to agree geographical spheres of influence for chemical sales. In particular the verbal battle over spheres of influence between W. S. Robinson and his great private friend and business antagonist, Lord Macgowan, Chairman of I.C.I. in the 'twenties, is one of the most amusing sections of W. S. Robinson's papers.

Progress in acid disposal for profit had, therefore, necessarily to be slow during the first twenty years of National Smelting's existence but this constituted no great problem as the amounts of acid for disposal were very limited from the end of the 1914–18 War until the start-up of the vertical retort plant in 1934. The hesitant start-up of the Avonmouth furnaces, followed by limitation of zinc production by the International Zinc Cartel until the end of 1934, were the main reasons for this. Disposal of acid from the Swansea Chamber plants remained the responsibility of the Briton Ferry Chemical & Manure Company, as explained earlier in the chapter, until 10 September 1924 when notice was sent to that Company 'terminating the existing arrangements'. There is no evidence that the ex-Government acid plant at Avonmouth produced any acid at all between the end of the 1914–18 War and the taking over of the manage-ment by the Australian interests in December 1923; in fact, there are clear indications that it remained closed.

On 18 February 1924, R. G. Perry, head of the National Sulphuric Acid Association, and Lord Wargrave of W. & H. M. Goulding, the Irish acid and fertilizer manufacturers, joined the Board and negotiations began on a con-tract with Perry's Association to dispose of the Company's acid from both

Avonmouth and Swansea Works. This was signed eventually in September 1924 and formed the main outlet for acid from 1924 to 1929, but no regularly maintained production and sales figures are available for these years to show what use was made of this outlet.

In 1924 also the idea of superphosphate fertilizer manufacture is mentioned for the first time and a report on this possibility was to be obtained from Gouldings; but in May 1925, after consideration of the report, it was decided 'that there was no inducement for the Company to further consider the question of entering upon this business at the present time' and the idea was not revived until 1934.

In June 1925 Perry reported a serious situation in the acid industry. Only 40 per cent of the country's available capacity was being used, compared with 65 per cent in May, and National Smelting's output at that moment represented 20 per cent* of the total. It was at moments like these that Perry was repeatedly advocating in these years the policy of engaging in subsidiary industries to absorb part of the Company's acid rather than getting rid of it on the market and thus depressing the price—echoes of the Government Committee's findings of 1918. W. S. Robinson opposed this policy with the view that the Company should first complete its smelting, roasting and acid construction programme before entering into new commitments of this nature. Finally, in February 1929, Perry's policy prevailed and 'The Managing Director was given power to take such steps as he thought fit to enter into negotiations with any organizations interested in the use of such subsidiary products as would be available to the Company if its works were extended'.

From 1930, after extensions at both Avonmouth and Swansea, the problems of acid disposal became more complex. Direct sale efforts centred round a new agreement with the National Sulphuric Acid Association and with six small acid producers in South Wales, which had been executed on 11 December 1930 and was to run for ten years. This agreement reflects again the forecast of the Government Committee of 1918 that many chamber plants would be redundant after the war and that the advent of acid from zinc concentrates would hasten this process.

Briefly the six small acid producers in South Wales (interpreted as extending up to Gloucester, West of the Severn) had formed in the late 'twenties an Association to reduce unrestricted competition. They were also members of the National Sulphuric Acid Association. Under an agreement between them and the newly formed Imperial Smelting Corporation, they undertook to close down their plants and give up their market to Imperial Smelting in

* It has now fallen to about 8–10 per cent.

return for compensation from Imperial Smelting on an annual basis which was to be paid for seven years until the end of 1937. The distribution agent was to be the National Sulphuric Acid Association. The records show that, over these seven years, 624,950 tons of Imperial Smelting acid were disposed of through this agreement at an average delivered price of £2. 11s. 6d. a ton and with a handsome profit both to the 'vendor companies' and to Imperial Smelting.

The main interest of these years, however, is the use that W. S. Robinson made of the powers given him by the Board on 18 February 1929 to negotiate with organizations interested in the use of our 'subsidiary products' (i.e. mainly acid) following the policy advocated by Perry.

The first venture was a modest one but was destined to lead the Company in the direction of fluorine chemical production, which has been the most important aspect of its chemical activities in the past decade.

On 20 September 1933 the Board gave its approval to a proposal to set up Aluminium Sulphate Limited. Obviously, from the wording of the Minutes, the matter had been under discussion for some months between W. S. Robinson and his friend, Murray Morrison, Managing Director and later Chairman of British Aluminium, and a very close association between the two companies developed which survived the failure of the joint magnesium venture to be described later in this book.

Aluminium sulphate is a product of purified aluminium hydrate and sulphuric acid. It is used widely in paper production and as a chemical in the textile dyeing industry. This joint venture with British Aluminium has been one of Imperial Smelting's steadiest, if least spectacular, investments and has continued to this day.

The original plan was for 70 per cent participation by British Aluminium and 30 per cent by National Smelting in initial capital of £30,000. Out of this a factory for production of 10,000 tons per annum of iron free aluminium sulphate was erected on the Avonmouth site on land leased from National Smelting. The plant started up in January 1934 and its consumption of National Smelting sulphuric acid was estimated at more than 4,200 tons a year. From the beginning National Smelting agreed to be responsible for production and The Alumina Company, a wholly owned subsidiary of British Aluminium with an aluminium sulphate works of its own at Widnes, for sales.

After less than a year's working the plant was already producing at 50 per cent above rated capacity to meet rising demand and was extended to 15,000 tons a year capacity in 1935. This decision was aided by an agreement with continental interests in May 1935 to restrict imports of their aluminium sulphate to 6,600 tons per annum. Production and sales for the last full year before the 1939–45 War were just over 13,000 tons with acid intake from National Smelting of about 6,256 tons. The net profit for the year ended 30 June 1938 was just under £11,000.

Participation in the equity capital was adjusted to a 50/50 basis between British Aluminium and Imperial Smelting in 1956. In 1960 Aluminium Sulphate took a third share in Thames Alum Limited, a new Company formed to set up a plant for production of aluminium sulphate in liquid form at Gravesend in the London area. It has various other ventures in hand.

From this aluminium sulphate venture started a significant development, which will be described in a later chapter. This was the joint work done by the research staffs of British Aluminium and Imperial Smelting on a process based on the reaction of hydrofluoric acid with aluminium hydrate for the production of aluminium fluoride, a product needed by British Aluminium for the production of aluminium and not then produced in Britain. The scheme put up involved installation of two separate sections, an aqueous hydrofluoric acid plant and the aluminium fluoride plant. Both plants were to be built on the basis of plans purchased from Herr Schuch who had 'designed and erected several plants in Germany'. British Aluminium agreed that the aqueous hydrofluoric acid plant should become the exclusive property of National Smelting, who might wish to develop the market for hydrofluoric acid on their own account, while the aluminium fluoride plant remained with Aluminium Sulphate Limited.

Little was done, except research work, to implement these proposals until 1939 and their importance belongs to a later period.

The second venture designed to provide an outlet for sulphuric acid in these years was Imperial Smelting's definite entry into the fertilizer business after several years of contemplation.

The first step was a modest one. One of the least inefficient members of the National Sulphuric Acid Association in South Wales was the Basic Slag & Phosphate Companies Limited, which was in itself an amalgamation of several small companies which had existed in South Wales at an earlier date. Imperial Smelting acquired this agglomeration from the shareholders (principally Baldwins, the steel producers) for £165,000 gross at the end of June 1934 and the acquisition brought with it besides 'waggons, engines, and stores' a miscellaneous collection of other small assets, including a factory at Newport which was at that time making, among other things, fertilizers from chamber plant acid and a little battery acid and distilled water. In addition, Basic Slag brought in small plants (on leased sites) for grinding into fertilizer the phosphate slag of various steel works. Those still working in 1934 were situated at Panteg, Gowerton and Port Talbot.

Later in the same year, 1934, the way was cleared for the second and much bigger involvement in the fertilizer industry by the conclusion of an agreement with Fison, Packard & Prentice Limited (this company became 'Fisons' in 1942) for a joint enterprise in fertilizers which was incorporated on 26 October 1934 under the name of National Fertilizers Ltd. The arrangements setting up this Company were even more complicated than those on which Basic Slag had originally been founded. Imperial Smelting put in the fertilizer and slag grinding plants and business of Basic Slag and Fisons, Packard & Prentice put in their interests in the Western and Merseyside areas, while retaining for themselves the bulk of their business in the Eastern and Home Counties. The original equity capital was £10,000 of Ordinary Shares subscribed in cash by Imperial Smelting and the main intention was to erect a new fertilizer works. This was financed initially by an issue in November 1934 of 200,000 5 per cent Non-Cumulative Preference Shares subscribed in equal proportions by Imperial Smelting and Fisons. The following year a further 50,000 Preference Shares plus 140,000 Ordinary Shares were issued, making a total capital of £400,000. By 1935 Fisons had supplied £204,000 (51 per cent) and Imperial Smelting/Basic Slag £196,000 (49 per cent) of the total capital.

The factory was erected on Imperial Smelting's Avonmouth site and managed by Imperial Smelting. It was connected with the dockside by a new ropeway to bring in imported phosphate rock as the Docks Authorities would not permit the fertilizer works to be built near foodstuffs warehouses on the docks, which was one of the possibilities originally considered.

The most important aspect of the whole deal, from Imperial Smelting's point of view, was an agreement to supply the new fertilizer works, as well as the small works taken over at Newport, with 44,000 tons of 66·52 per cent strength acid for a period of fifteen years at an agreed price. This was the largest sale of acid yet made by Imperial Smelting and eased the acid surplus position greatly after the increase in roasting of sulphide concentrates at Avonmouth brought about by the start-up of the vertical retort plant.

The new fertilizer factory at Avonmouth started up in July 1936 but by this time the scope of the business had been increased by further deals which, to pass over them briefly, comprised a takeover of the superphosphate works of Charles Norrington at Plymouth, of the fertilizer business in the Western areas of England of Anglo-Continental Guano Works, of a part interest in George Hadfield of Liverpool, and of the whole business of Thomas Vickers & Sons of

Widnes. These interests brought the turnover of National Fertilizers up to 45,000 tons of superphosphate fertilizer a year by mid-1936 which necessitated an expansion of share capital to £400,000 worth of Ordinary Shares and 500,000 4¼ per cent Cumulative Preference Shares (mainly owned by the public) in September 1936. The partners also agreed to form a subsidiary (Corby Basic Slag) with capital of £75,000 to acquire a new slag grinding plant being erected by Stewarts & Lloyds, the steel producers, at Corby, with the right to fifteen years' slag. This plant was soon producing an average of 9,000–10,000 tons of ground slag regularly each month whereas production from the three surviving plants of Basic Slag never exceeded 1,000–2,000 tons altogether.

The net profits of National Fertilizers rose from £7,896 in 1935 to £42,575 in 1939 by which time Newport Works was producing 300–400 tons of special fertilizers a month to supplement Avonmouth's production of straight super-phosphate.

The most obvious outlet for sulphuric acid is in fertilizer manufacture and the decision of Imperial Smelting to give up a direct interest in the fertilizer company that it had helped to create in 1934 has seemed strange to the present management of the Company at this distance of time.

The reasons, as given by L. B. Robinson to the Board on 2 October 1942, were that 'the two companies concerned and their subsidiaries had adopted their own selling methods under their respective brands and the existing circumstances presented an opportunity of centralizing the selling arrangements under National brands and of launching advertising campaigns'. Also I.C.I. were the only producers of concentrated fertilizers at that time and National Fertilizers and Fisons, Packard & Prentice were not large enough to enter this market and survive individually. Undoubtedly also an underlying complication was that, by 1942, Fisons, Packard & Prentice were paying a dividend of 10 per cent whereas National Fertilizers were paying only half that percentage and clearly National Fertilizers would have fared even worse if Fisons had decided to sell their own brand on a national basis. Imperial Smelting, therefore, found little difficulty in agreeing to accept a 16½ per cent interest in the whole of Fisons growing profits, swollen by those of National Fertilizers, and two seats on Fisons Board, in place of the frustration of their fertilizer ambitions in the anti-climax of National Fertilizers' actual results. These ambitions were, however, to haunt the Boardroom as a possible panacea whenever a sulphuric acid surplus appeared in the next thirty years and to take modified terrestrial form again in the recent decision to produce phosphoric acid to absorb some of the acid from the new Avonmouth expansion scheme.

The only other chemical venture of these years which was established on a commercial scale in Britain by Imperial Smelting is worthy of mention, although it had no connection with sulphuric acid and only a very remote connection with zinc through the use of zinc naphthenate as one of the ingredients of the product.

In 1932 Stanley Robson and Dr. Stacey Lewis, who later left the Company's service and took Holy Orders, went to Copenhagen and negotiated the acquisition from a Danish Company of the U.S.A. and British Empire rights to a patented process for inhibiting rot in fishing nets, woodworm, dry-rot and allied afflictions to which wood and comparable substances are subject. A Company was set up under the name of Cuprinol Limited in which National Smelting received a controlling 51 per cent interest in return for cash (raised by the sale of 13,000 more Burma shares at 10/10d. each) and the Danish Company a/s Kymeia received the remaining 49 per cent in return for the patent rights. A small factory to produce 100 tons a year was erected on the edge of the Avonmouth Works site and started production at the end of 1933. This was the type of product which requires a widespread advertising campaign and a network of stockists and travelling representatives for success and, to judge from the sales reports of those days, these were all provided without stint. A retired colonel had a great deal to do with organizing the sales campaign and the reports are full of invasions of new territory with Cuprinol, including old churches, canvas, agricultural shows, Scotland, and Windsor Castle. Gallons ever increasing in number were made and sold inside Britain and outside. But somehow a profit never appeared in the annual accounts before the 1939–45 War and in 1937 W. S. Robinson wrote that 'the business of Cuprinol shows some but not sufficient improvement to justify the belief that both the production and sales side can yet be run by our organization successfully'. Jenson & Nicholson, the paint manufacturers, were invited to take over distribution in Britain and later on, after the war, became 49 per cent partners in it with Imperial Smelting. In 1958 they acquired Imperial Smelting's 51 per cent interest and production of Cuprinol ceased on the Avonmouth site shortly afterwards. The heyday of the product in its multiple varieties was during the 1939–45 War when it was widely used in the treatment of sandbags and it has had increasing success at the present day as a product produced and sold by Berger, Jenson & Nicholson Ltd.

Further expansion into chemicals was temporarily halted by the war but was resumed more systematically afterwards.

CHAPTER ELEVEN

1929-1939
Sidelines

DIVERSIFICATION IS RARELY LOGICAL—FOUR INTERESTING
VENTURES OF THE 'THIRTIES—LITHOPONE AND THE
ACQUISITION OF ORR'S ZINC WHITE LIMITED
ITS RELEVANCE TO ZINC—BARYTES MINES
PROGRESS AND NEW VARIETIES OF PIGMENT IN THE 'THIRTIES
THE INTERNATIONAL LITHOPONE CARTEL ITS BREAK-UP
AND LATER COMPETITION FROM CONTINENTAL IMPORTS
THE DISCOVERY OF TITANIUM DIOXIDE PIGMENTS AND THEIR
DEVELOPMENT BY NATIONAL LEAD COMPANY IN THE U.S.A.
DR. JEBSEN AND THE EXPLOITATION OF THE EUROPEAN AND
EMPIRE RIGHTS—FORMATION OF BRITISH TITAN PRODUCTS
THE DECISION TO ESTABLISH WORKS AT BILLINGHAM INSTEAD OF
AVONMOUTH—SPECTACULAR LATER PROGRESS
ZINC OXIDE AND THE ACQUISITION OF A MAJORITY INTEREST
IN FRICKER'S METAL COMPANY—PROLONGED NEGOTIATIONS
COMPETITION AND THE FAILURE OF EFFORTS TO FORM A CARTEL
LATER VARIETIES OF ZINC OXIDE—ZINC DUST FOR USE IN GOLD
PRECIPITATION—KEEPING UP WITH GERMANY AND THE
MAGNESIUM VENTURE—COMBINES, INVOLVEMENT AND PROCESSES
THE HANSGIRG PROCESS—IMPERIAL MAGNESIUM SETS UP A
PLANT AT SWANSEA—OPERATIONAL EXPLOSIONS AND DIFFICULTIES
OVERSUPPLY AFTER THE WAR
AND THE CLOSURE OF THE SWANSEA PLANT

IN ALL PROBABILITY it is impossible to write a history of any large industry which can claim to unravel a single logical thread of development over the years and still remain true to the facts. Human beings are rarely logical and the structure of Company Law in Britain still gives unusual opportunities to those whom it channels into the chief seats of industry, to try out their ideas without having to face opposition on equal terms from their shareholders or their immediate entourage. As a Research Director of Imperial Smelting was once informed by a fellow scientist 'The main function of the Research Director is to shield his staff from the occasional flashes of genius emanating from the Managing Director!'

It would be untrue, therefore, to pretend that four important activities of Imperial Smelting to be outlined in this chapter were an inevitable outcome of the intention to produce zinc metal in Britain from Australian concentrates, although two of them had a direct connection with zinc and were the result of a growing interest in developing all profitable outlets for zinc and its derivatives, more especially in pigment uses. The third was a brilliant inspiration which, with the other two pigment ventures, has played its part in bolstering up Imperial Smelting's fortunes at times when, owing to depressed prices, zinc smelting must scarcely have been worth continuing as an activity on its own. The fourth was an energetic incursion into a segment of the non-ferrous metal field remote from zinc and became a disastrous but interesting failure.

To deal with the more successful projects first it will be remembered that mention was made in the oft quoted Annual General Meeting of 31 July 1929, at which the proposed formation of Imperial Smelting was first publicized, of 'a special interest in developing the production of lithopone and zinc oxide'. On 18 September, W. S. Robinson, in a letter to the Board, talked of the possibility of 'acquiring New Jersey's lithopone patents as well as those of the vertical retort, zinc oxide and metallic zinc alloys'. He went on to say that the fairest and, probably, the easiest course to adopt appeared to be 'the acquisition of efficient existing interests if such are obtainable at a reasonable price. Negotiations have been opened through Sir William McClintock, partner in a firm of Chartered Accountants with this in view.' On 16 January 1930, his son, L. B. Robinson, submitted a detailed recommendation to the Board for the purchase of the whole of the share capital of Orr's Zinc White Limited, a private company owned almost entirely by J. B. Orr, the founder, Donald J. W. Orr, his son, W. A. Reid 'and their direct family connections'. The purchase would bring with it also control of the Wrentnall Baryta Company which owned a barytes mine in Shropshire.

The recommendation was accepted and the deal went through with great rapidity in February 1930. The ledger figure given for the purchase price for the whole of the share capital of the two companies is £498,799.

The famous old landmark of Widnes, Lancashire, the Vine Works of Orr's Zinc White, had a long and interesting history prior to its acquisition by Imperial Smelting. The Company had been incorporated in 1898 and the founder of the business, John Bryson Orr, was born in Lanark as long ago as 1840. The relevance of the business to Imperial Smelting's zinc activities was that the pigment lithopone, which was its main product at this date, is 'primarily an intimate chemical mixture of zinc sulphide and barium sulphate giving a dead white product'. This product is used as a pigment and also as a 'filler' in the linoleum and rubber trades and Orr's Zinc White Limited, which was renamed Barium Chemicals Limited in 1964, ceased producing it only in that year.

As regards the zinc content of lithopone, it was stated at the time of purchase of Orr's Zinc White that 'any zinc bearing material may be used in the manufacture of this product, and although National Smelting could supply roasted zinc concentrates to Widnes for this purpose, it may be found that suitable material such as zinc ashes can be drawn from sources nearer at hand'. Actually, throughout much of the 'thirties Orr's depended on Canadian (Buchan's) zinc concentrates roasted on a Delplace roaster at Seaton Carew Works for the zinc sulphide side of lithopone.

The barium sulphate side came from barytes supplied from the Wrentnall mine, supplemented by purchases from Germany and from the Devonshire Baryta Company. Later on in the Company's history Wrentnall became worked out and was succeeded in 1937 by Cow Green Mine, County Durham, by Gasswater in Ayrshire and by a part-interest in Muirshiel mine sometime in the mid-thirties. The operation of these small mines absorbed possibly a disproportionate amount of the Imperial Smelting's managerial effort, although they frequently provided a pleasant day in the country for 'chairborne Directors', and it is not possible to describe their history in detail within the compass of this book. Outside sources of supply also altered with the years and, in recent years, imports of barytes from Ireland, Morocco, and Spain have become increasingly necessary. A complicating factor was that the Company's other principal product was ground white barytes, which is used as an extender in the paint and allied industries and requires a high grade of untinted white barytes originally obtained from Gasswater.

Orr's Zinc White, at the time of the takeover, was a flourishing business, as it was then the heyday of lithopone, and the management was left largely in the hands of its previous owners. Many of the employees have been retiring in

I

recent years, some after as much as fifty years' service. Annual production at Orr's at the time of takeover amounted to about 17,000 tons of lithopone, being the maximum permitted by the International Lithopone Union (see later), and 6,000 tons of ground white barytes. Pre-tax profits for the year ended 31 July 1928 were £34,900 and £44,600 for the following year. The Company had a twenty acre site and most sections of the works were, in 1930, in a fairly good state of repair with the exception of the North Plant. A programme of expansion was at once put in hand to expand the other (Lugsdale) finishing plant and to construct a third production unit, which was completed in 1932, and a new dryer.

Sales dropped to only 13,754 tons of lithopone in the depressed year of 1931 but rose again to 17,807 tons in 1932 and to 34,000 tons by 1938. With Protection in 1932 came a windfall in the shape of a protective tariff of 10 per cent against foreign lithopone which, with the production economies made possible by the expansion of the works, enabled Orr's to reduce the price of the product.

Meanwhile, a good deal of research and management effort was being put into developing new qualities of lithopone and new zinc sulphide pigments. On 15 December 1932, L. B. Robinson reported that a licence had been obtained from I.G. Farbenindustrie to operate that Company's cobalt process for lightproofing 'and this has offered a further field for the development of a range of qualities for which there is a market in this country. Apart from high strength lithopone (60 per cent zinc sulphide) three new qualities of the 30 per cent grade have reached the bulk trial stage . . . In addition to the above, work is being conducted towards the production of a titanated lithopone—a mixture of titanium and lithopone—and also pure zinc sulphide. At present the market for these products in this country is extremely small, but in the case of zinc sulphide, recent developments in the U.S.A. indicate the possibility of quite a substantial trade being developed.' This plan led to further expansion of capacity in 1933. A fourth production unit was built to produce 60 per cent lithopone (i.e. roughly 60 per cent zinc sulphide and 40 per cent barium sulphate) and a patent covering production of 100 per cent zinc sulphide was licensed from The New Jersey Zinc Company while work on a half-unit to produce this was expedited. It was thought that pure zinc sulphide would be capable of competing with titanium dioxide in the pigments market but this proved a forlorn hope.

The overall result of these efforts was an increase in the plant capacity for zinc sulphide pigments to over 30,000 tons a year and a steady profit rising from £35,468 in 1931 to £62,500 in the year ending 30 June 1939. The annual payment of most of this profit to the owners largely enabled Imperial Smelting to pay its Preference dividends in these years when zinc smelting was virtually unprofitable.

Before leaving the subject of Orr's Zinc White, two interesting aspects of this business should be mentioned. W. S. Robinson had the New Jersey patents and process for lithopone manufacture in mind when he urged the Board to enter the lithopone business, but it seems obvious that no use was made of the New Jersey patents even if they were ever acquired.* It is stated in a letter by Stanley Robson dated 9 January 1930 that lithopone was first manufactured by a Glasgow chemist, John Bryson Orr (1840–1932) the founder of Orr's Zinc White Ltd., and was first called 'Orr's Zinc White' and not lithopone, even in Germany. 'All other manufacturers have followed the general lines of his inventions—I am told that many paint and linoleum manufacturers will purchase only this brand and that the New Jersey Zinc Company's brands possess different physical characteristics. . . .'

Another interesting aspect of this business in these years is that Orr's Zinc White Ltd. brought with it into Imperial Smelting obligations, as a member of an international cartel, almost as irksome as those binding National Smelting to the Zinc Cartel of this period. It was participating in 1930, in a Convention Agreement between the lithopone interests in Holland, France, Belgium and Germany, its share at that time being $12\frac{1}{2}$ per cent of the total convention production of approximately 140,000 tons per annum. This was, of course, the epoch between the encouragement of industrial cartels by the German Empire before the First World War and their outlawing by the European Economic Community in the late 'fifties. They were fragile inventions even in those days as, having grown up in an atmosphere of free trade and unrestricted competition, they were usually unable to cope with the problem of the imposition of national tariffs, protecting one or more of their members. When the possibility of Britain adopting Neville Chamberlain's protection proposals became increasingly likely in 1932 the first reaction of the International Lithopone Union members was to consider a scheme for utilizing Widnes Works as a Convention Works to overcome the problem of tariffs being imposed on lithopone imported into Britain from the Continent. The idea was that lithopone would be manufactured to their standards at Widnes to enable them to maintain their sales within the British Empire at existing levels. Representatives of Sachleben and I.G. Farbenindustrie visited Widnes in 1932. L. B. Robinson made a return visit to Germany and installation of a third unit was put in hand to increase production at Widnes. Agreement on supplying the Empire from Widnes was near in June 1932, but was defeated by the intervention of the German, Dutch

*No legal document or other reference to a transfer is traceable.

and Belgian Governments who 'effectively prevented' the proposed international arrangement for the interchange of patents and processes and the manufacture in Britain of 'the German, Dutch and Belgian lithopone qualities'.

At a meeting between the Convention members on 12 September 1932, 'it was decided that all parties were now free' and as W. S. Robinson concluded, 'for the first time for many years Orr's Zinc White is now able to bid for world trade'.

The break up of the Convention in 1932, however, meant that competition from imported lithopone continued throughout the 'thirties. By the end of 1936 imports were averaging 1,200 tons a month and increased steadily up to 1939. Prices had to be reduced to retain business with the linoleum and rubber trades in the face of continental competition and, when joint application was made by McKechnie Brothers Limited, the only other producer in Britain, and Orr's, early in 1938, the import tariff was increased from 10 per cent to £3. 5s. od. a ton or 20 per cent *ad valorem* (whichever was higher). This increase appears, however, to have had little effect on reducing imports, the suspected reason being that continental producers were 'dumping'.

Lithopone production continued at Widnes until early in 1964 when it was closed down in the face of a steady decline in demand since the late 'fifties. Orr's was still a member of an international association at the end but the purpose of the association was, by that time, to prevent dumping and unreasonable competition.

Lithopone declined in the 'fifties because of the increasing popularity of another pigment, titanium dioxide, which developed technical and price advantages over it.

National Smelting probably became involved in this development in 1927 when the first reference to titanium dioxide pigments appears in the Minute Book, but the history of the development of these pigments began a long time before then.

The traditional story, which no one now wishes to question, is that the Reverend William Gregor, a Cornish parson, painter, musician and amateur chemist, was the first to isolate the white oxide of a new element, titanium, while examining the composition of a black sand found on the beach at Manaccan in Cornwall early in the nineteenth century. The legend does not say whether he made any commercial use of his discovery.

There is very little evidence of any further development until the first decade of the present century when Dr. Gustav Jebsen started his experiments to produce titanium dioxide from Norwegian ilmenite deposits, but there is a detailed account of events since then in documents submitted to two U.S.A. courts in 1945-46 in a case, instigated under the Sherman Acts, to end alleged restrictive practices by the titanium pigment companies. Some of these documents have been lent to the authors through the kindness of Mr. J. Askew, Secretary of British Titan Products. They are interesting from Imperial Smelting's point of view as the story is current that W. S. Robinson was mainly responsible for initiating the commercial exploitation of the Empire rights to Jebsen's patents which led to the formation of British Titan Products, the British company which is now one of the three or four world leaders in this field. These U.S.A. documents show quite clearly that Jebsen took his inventions to the U.S.A. immediately after the 1914–18 War and that, by 1927, the National Lead Company owned the world and Empire rights in Jebsen's inventions. Jebsen also seems to have become an employee of National Lead about this period and certainly had been actively associated with them before that through an agreement made between National Pigments (later acquired by National Lead) and his Norwegian Company in 1920.

W. S. Robinson's statement, therefore, in his letter to G. W. Beeby, present Chairman of British Titan Products, written in 1961, that he went over to Norway to meet Jebsen and inspect his product and then, after first failing to interest the National Smelting Board in it, went over to the U.S.A. to get National Lead support to a British Empire enterprise, seems a little one-sided. According to the U.S.A. documents 'Jebsen had for years attempted to establish a British market for his pigments but with limited success'. Certainly, a prominent executive of I.C.I. of that period, Colonel J. S. Barley, when he wrote down his memories of these early years, also in 1961, contended that he had first met Jebsen in 1923 and expressed the willingness of Nobel Industries Limited (which in 1926 was merged in the new Group, I.C.I.) to take up the commercialization of Jebsen's patents whenever Jebsen thought the time was suitable. When Barley met Jebsen again in 1929 Jebsen explained that he had had an approach from National Smelting who wished to find an outlet for some of the acid resulting from their smelting operations.

The account of Jebsen's position in the 'twenties* as given in the U.S.A. documents fills in some of the background to these negotiations:

---

* It is interesting but rather irrelevant to the present account to note that a titanium pigment company was already in existence in Britain in the 'twenties. The U.S.A. documents say: 'British Titan was not the only manufacturer of titanium pigments in the British Empire. A small company, National Titanium Pigments Limited, known as Laporte, having acquired rights under the so-called Tilden Smith patents and under the Blumenfeld patents for the British Empire, entered into agreement with British Titan fixing prices in the British market excluding Canada.' Laporte, of course, are still producing titanium pigments and yet a further reference to the activities of Tilden Smith, who features so largely earlier in this book, will not have escaped the reader's attention!

Knowing from costly experience that he could not hope successfully to invade, much less to capture, the British Market without a strong local alliance, Jebsen set out to obtain one. Though I.C.I. had never manufactured pigments, it was, obviously, the most desirable British partner, for it had resources in the way of technical and chemical facilities, raw materials, etc. not possessed by any other firm. Imperial Smelting Corporation Ltd. and Goodlass Wall and Lead Industries Ltd. were also natural allies. I.S.C. (i.e. Imperial Smelting) was the largest British manufacturer of zinc and lithopone: it could supply sulphuric acid; it had important connections throughout the Empire; and it had strong management friendly to National (Lead). Goodlass Wall, a government sponsored amalgamation of lead companies, was brought in for similar reasons.

According to Barley, Goodlass Wall were also alive to the threat of titanium dioxide to white lead pigments and had been trying to evolve a process of their own.

British Titan Products are contemplating producing a book on the history of their company and the full and immensely complicated detail of the negotiations of the 'twenties will doubtless be contained in this. Sufficient, however, has been said here to indicate that the truth is probably that it was in the best interests of National Lead, Imperial Smelting, I.C.I., Goodlass Wall and R. W. Greef (formerly Jebsen's sales agent in England) to come to some sort of arrangement about the exploitation of the British Empire rights. The position, as appears in the Imperial Smelting Board papers of 22 September 1932, was that National Lead had, by that time, extensive titanium oxide factories in the United States but that the whole of its European business, including the British market, was supplied from a plant operated by I.G. Farbenindustrie at Leverkusen in Germany.

The proposals, which were eventually embodied in agreements, were that British Titan Products Ltd., a small 'dummy' company already formed in anticipation by Imperial Smelting in 1930, should be taken over by the parties to the main agreement. Then, as from 1 April 1933, British Titan Products would take over the National Lead Company's business in titanium products in the British Empire with the exception of Canada, but with the right also to take over the benefit of the I.G. Farbenindustrie contract to supply the British market with the I.G. Farben product on very favourable terms until a plant had been built in Britain.

Initial capital of £125,000 was put up by National Lead (49 per cent), I.C.I. (17 per cent), Goodlass Wall (17 per cent) and Imperial Smelting (17 per cent) for the acquisition of the business and all its patents and trimmings and the profits were to be shared in the same proportion. R. W. Greef & Company Ltd. took over 5 per cent from National Lead's share shortly afterwards.

The Imperial Smelting Board, which had at first been very doubtful about this project, appears to have been won over when National Lead said that they would support this scheme.

The original idea was that the new British plant would be 'housed with I.S.C.' and located at either Avonmouth or Widnes. A Norwegian engineer and chemist were appointed. A technical committee was also appointed, among other things, to choose a site and, although it at first favoured Widnes and the Imperial Smelting Board strongly favoured Avonmouth, later chose Billingham for the site which was subsequently purchased from I.C.I.

Billingham was a good site and easily accessible to raw materials from Norway via the Tees. Apart from this disagreement over sites the remaining history of relations between the companies who have made up 'B.T.P.' has been unusually harmonious and the support of I.C.I., who have given some of their best men to the top positions, has been particularly welcome.

The British business in 1933 was reported to be yielding about £8,000–£10,000 per annum profit after allowing for duty and exchange and the first objective was to build the plant at Billingham. This was estimated to cost £147,780 which was raised by issue of a further £175,000 worth of shares by B.T.P. It started up in July 1934 and expansion of capacity became necessary almost from the moment of start up.

Although it has always been regarded as a single product Company there were, from the beginning, two grades of pigment, known as rutile and anatase from the crystalline forms of the natural mineral, and new grades for special uses gradually multiplied. Its original use was in paints but it is now used in a large variety of products. A chloride process, which was developed outside Du Pont's patents in this field, will probably supersede the original sulphate process and in the meantime has enabled British Titan Products to become the only Company to sell titanium pigment produced by the chloride process apart from Du Pont.

Accordingly, growth, particularly since the 1939–45 War, has been a success story of the most spectacular type.

A further plant was built at Grimsby in 1949 and later on plants were built in Australia, Canada and South Africa. B.T.P. has also acquired an interest in an Indian plant and a new plant is under construction at Calais inside the Common Market area.

The net profits, which in 1938 were stated to be £74,000, were over £2½ million in 1965, and the capitalization nearly £32 million. Tremendous effort has, of course, been required to establish this new industry in Britain and, naturally enough, growth to the present dimensions did not really start until after the war years. The delay in expansion was unlucky for the Company to which B.T.P. owes so much, The National Lead Company, as they lost their appeal to the U.S.A. Supreme Court in 1946 and were compelled to divest themselves of the remainder of their interest in B.T.P. Their shareholding was divided out among the other shareholders and Imperial Smelting, I.C.I., and Goodlass Wall now own a little over 30 per cent each with R. W. Greef & Company holding just under 9 per cent.

This investment has indeed provided Imperial Smelting with almost fairy tale riches over the years compared with the hard struggle to wring a few pounds profit out of zinc.

In 1930, when outlining the project, W. S. Robinson had informed the Imperial Smelting Board that 'There is no (capital) commitment beyond, say, £3,400 of an original £20,000 unless we are satisfied to go on'!

The third pigment project initiated in these years was zinc oxide, used in those days primarily in paints and increasingly in rubber and tyre manufacture but now mostly for tyres.

Investigations into the possibility of taking over Fricker's Metal Company as a quick and convenient method of entering the zinc oxide industry, began in 1929 but the first report to the Board, that can be traced, is dated 14 May 1930. This shows that at that time Fricker's Metal Company had an issued share capital of £88,000 and two small works, one at Luton and one at Burry Port. It was stated in September 1932 that 'The Company has been taking for some time past between 15 per cent and 20 per cent of National Smelting's spelter output at a satisfactory premium and profit to us'. Luton produced 3,557 tons of zinc oxide in 1928 and 4,715 tons in 1929 and Burry Port 1,710 and 2,190 tons respectively in those years. Average annual net profit for the two years was £16,155. It was claimed that 'prior to the year 1928 the Company had been earning little if any profit and the marked improvement shown from 1928 is accounted for by a change in the process of manufacture brought about during the latter half of 1927 on the appointment of Mr. Lindsay Scott as Managing Director'.

Imperial Smelting were prepared to offer a price that they considered generous because in 1930 the annual consumption of zinc oxide in Britain was estimated to be approximately 30,000 tons of which some 12,000–14,000 tons was imported from the Continent and America. It was hoped that, by expansion of Fricker's plant, the business would be placed in a useful competitive position to capture at least the portion of the market met by imports. If a new works were built forthwith on a new site time would be against the enterprise. Also 'the task of marketing a new brand of product in the pigment trade is likely to

Automated Bell charging hoppers at Imperial Smelting Furnace, Swansea

Control room of the sulphuric acid plant at Avonmouth Works of Imperial Smelting

Tapping slag from an Imperial Smelting furnace

be a long one; a paint manufacturer has his formulas adapted to a particular brand, and he cannot risk a change without exhaustive tests'. This is a well-known factor in the pigments industry and one of the main reasons for acquiring Fricker's rather than setting up a new works. Another reason was that Luton was attractively near London, which is a major centre of the paint trade, and 'as well if not better situated to serve other centres than, for instance, Avonmouth'.

Two more years went by, however, before Imperial Smelting acquired a majority interest in Fricker's. The Board decided to postpone the purchase 'in view of the distributive trading situation', but in mid-April 1931 the threat of foreign competition in the British zinc oxide market prompted it to reopen negotiations. Although it was proving difficult to raise a loan for the proposed purchase owing to the financial situation in London, the possibility of zinc oxide receiving tariff protection appears to have expedited the matter and the terms on which Imperial Smelting would take a 61 per cent interest in a new Company, Fricker's Metal and Chemical Company, were agreed finally with the management and shareholders of Fricker's in May 1932.

The total cost in cash to Imperial Smelting was £52,350, raised by the sale of 100,000 Burma shares and, although the remaining equity interest did not pass to Imperial Smelting until 1952, the negotiations of 1930–32 brought to it full control of Fricker's operations and sales and also the services of Mr. Lindsay Scott and the right to supply Fricker's with the special de-leaded sinter that it required for raw material. Again the familiar words appear, as in the case of other ventures taken over in these years—'arrangements are also being made for the probable utilization of buildings and site at Avonmouth' and again this intention to centralize operations at one site came to nothing.

The bulk of the shares in Fricker's were transferred, along with the shares in Orr's and 'B.T.P.', to Non-Ferrous Metal Products Limited, a specially created subsidiary company of Imperial Smelting. The idea which motivated this policy appears in one of the earlier reports to the Board on the 'final' terms for the acquisition of Fricker's—'the commercial advantages of direct linking up of spelter, zinc oxide and lithopone works are substantial'. Beyond this was a still larger vision of the conclusion of negotiations with 'our American friends' for a close association in 'zinc, lead and titanium pigments and other associated products' but this also came to nothing.

Actual results from Fricker's proved disappointing owing to price competition, particularly in the grades brought by rubber and paint producers. This was even admitted several times in the Chairman's annual speech during the 'thirties. Zinc oxide, unlike zinc metal, has always been a comparatively easy product for smaller 'backyard' works to make and the history of the past thirty-five years

A laboratory

has shown perpetual alternations between price warfare and attempts to form some sort of cartel to stabilize prices at a reasonably profitable level.

The first instance of this is recorded in 1932. By the end of that year competition was apparent, particularly from the new plant at Barking of M. Pisart who had substantial pigment interests in Western Europe. Sulphide Corporation had also brought a zinc oxide plant into production at Seaton Carew and Delaville were producing at Bloxwich. Accordingly, a Mr. Hutton of Eaglescliffe Chemical Company, which subsequently went out of production, came to Lindsay Scott at Fricker's with the first recorded scheme for a producers' agreement among the eight principal producers at that time. Lindsay Scott was unimpressed and the reply was given that, while Fricker's would support any sound scheme that could be put forward, it was doubtful whether a scheme could be devised to meet all recognized producers' points of view and at the same time avoid encouraging 'pirates'. Similar problems were to occur in later years and remain unresolved.

The bulk of Fricker's business at this time consisted of this production of an indirect process grade of zinc oxide from melting slab zinc, but in May 1933 a new product was introduced when a furnace for the production of 'direct process leaded zinc oxide' was installed at Luton principally in response to demand from the Canadian market. This was a form of co-fumed zinc oxide, produced direct from the reduction of special zinc/lead residuals rather than from slab metal, and very substantial tonnages of this product were then being used in the U.S.A. and on the Continent. By May 1934 Fricker's received its first two orders for this material, from paint manufacturers, and a new production unit was brought into operation at Luton. A further plant, to produce low leaded oxides, was installed at Burry Port in September 1935 and a second plant in 1938. By that time these direct process grades of zinc oxide were forming an increasing, although still minor, proportion of the Company's business. At the present day operations at Burry Port are confined to Indirect and Direct grades, production of leaded grades having been discontinued many years ago.

Another new departure of these years was also destined to endure. In December 1932 the Board were told that 'The production of zinc dust, which was recently commenced at Luton on a small scale, has been highly successful, so far as the quality of the product is concerned. But it is doubtful how far this product, manufactured from spelter, will be able to compete with the by-product material for the larger industrial uses.' The main outlet for this specially fine quality material was for use in the Merrill-Crowe process of gold precipitation in South Africa and many hundreds of tons were shipped there prior to the outbreak of the 1939–45 War. Subsequently a plant was erected in South Africa by Anglo American Corporation using the 'Fricker's' process. The product was also directed towards a totally different market from the coarser Delaville zinc dust as it 'appears very suitable for the manufacture of paints with excellent protective qualities for structural steel work etc.'. This high grade product has continued to be manufactured for this particular market but, on the closure of the Luton plant, the operation was transferred to the Bloxwich Works of Imperial Smelting where the name 'Fricker's' has been replaced with 'Delaville' for the superfine and ultrafine grades used in the manufacture of zinc rich paints. These are an extension of the original zinc dust paints, which contained a proportion of zinc oxide, whereas the zinc rich paints use only zinc dust as the pigment.

To conclude this chapter brief mention must be made here of the magnesium venture of these and the subsequent war years. This venture was of national importance but had no connection with zinc or sulphuric acid and proved a failure for several reasons not entirely the fault of Imperial Smelting.

In the decade from 1930 to 1940 interest in magnesium as a structural metal was growing, particularly in the Germany of that day which was striving for self-sufficiency. She had no domestic sources of alumina and looked to magnesium (magnesia being available from sea water even if there are no mineral sources) as a partial replacement for aluminium particularly in aircraft, for which its greater lightness was an advantage. I.G. Farbenindustrie A.G. had developed an electrolytic process involving the electrolysis of molten magnesium chloride and much research into the technology of the metal had been undertaken and application made to extend the patents interrupted by the 1914–18 War. The future of the metal looked rosy and it seemed likely that Britain was being left behind.

At this time (i.e. in the first half of the 'thirties) two companies in Britain were experimenting with magnesium production, The National Smelting Company through Calloy Limited and Murex through their subsidiary Magnesium Metals and Alloys Ltd. using reduction of magnesium oxide with calcium carbide. Two other major companies were vitally interested in magnesium production— Imperial Chemical Industries, because of their interest in chlorine, and British Aluminium, because of the possible effect on the market for aluminium.

In November 1935 these four companies agreed to form Imperial Magnesium Corporation Ltd. with a capital of £200,000, half in £1 Deferred Ordinary Shares and half in 5 per cent Preferred Ordinary Shares. The first were equally split between the four companies, and the second credited as fully paid in consideration of patents and processes as follows: I.C.I.—10 per cent; M.M.A.L. —40 per cent; B.A.C.—$17\frac{1}{2}$ per cent; N.S.C.—$32\frac{1}{2}$ per cent. National Smelting's interest was to be held through a subsidiary, Magnesium Metal Corporation, in which Calloy was to hold 20 per cent.

A technical committee of four was to be formed representing all parties, and their duty was to examine all possible processes including the Hansgirg process operated on a pilot scale at Radenthein in Austria by the Osterr.-Amerik. Magnesit A.G. (OAMA), which was well known for magnesite refractories. This process was based on the reduction of magnesia by carbon in an arc furnace at a very high temperature of over 2,000°C, the resulting magnesium vapour being shock-chilled with hydrogen and giving a fine dust which then had to be converted into ingot metal. A sub-committee spent three months in Radenthein examining the process which was actually owned by American Magnesium Metal Corporation through their Austrian subsidiary. However, before they had written their report and before any decision on the process had been taken, the unexpected happened. A company, later to become Magnesium Elektron Ltd. (48 per cent I.C.I. and 52 per cent Hughes and I.G. Farben) had negotiated with I.G. Farbenindustrie to use the electrolytic process in a plant to be built at Clifton Junction, Manchester.

In July 1936, according to the Board records, a difference of opinion occurred 'as to the respective merits of certain processes for the production of magnesium then under investigation by Imperial Magnesium Corporation Ltd'. As a result of this, in November 1936 I.C.I. and Magnesium Metals and Alloys withdrew from Imperial Magnesium Corporation.

Although perhaps a little hard on the other partners, this seemed a logical step for I.C.I., a large producer of chlorine, to take as they would prefer to back a process known to use chlorine rather than others which did not. It meant, however, that National Smelting and British Aluminium were left holding the fractious infant, Imperial Magnesium Corporation.

There followed a number of complicated company manoeuvres and changes of name which appear hopelessly over-elaborate in retrospect. The final outcome was the emergence in May 1937, of Magnesium Metal Corporation Ltd. (formerly American Magnesium Metals Corporation Ltd.) as the operating company, wholly owned by Magnesium Holdings (formerly Magnesium Metals Corporation), of which the equity capital was held by National Smelting and British Aluminium!

It would appear that these two had decided that the Radenthein process was the one most meriting serious attention and, immediately after the break, Stanley Robson himself went out to Radenthein and came back with a specific recommendation that only a pilot plant be built in England as the process was not an easy one and a number of snags could arise. Although he was later to be proved right, this recommendation did not suit Erdemann, the chief man at

Radenthein, who argued that they had already built a pilot plant, that Robson had overstated the difficulties and that there was nothing to stop Imperial Magnesium Corporation building a commercial scale plant (which would presumably put more money into the pockets of the Austrian Company). Unfortunately, Imperial Magnesium let themselves be persuaded and a plant was eventually built on part of the old English Crown Spelter site at Port Tennant, Swansea, by Magnesium Metal Corporation.

But troubles were only just beginning. The early stages of the process involve the handling of large quantities of magnesium dust. As the dust takes fire in air it must be transported in an inert gas—in this case hydrogen as nitrogen reacts with magnesium—and even a small leak of air or moisture into the system can result in an explosion. The sequence of operations is that the dust from the reduction furnace is rapidly chilled by hydrogen, allowed to settle and cool in a large horizontal drum cooler, and taken by screw conveyors to the briquetting presses whence the briquettes pass to the vacuum furnaces where the metal is distilled off and condensed as 'muffs' prior to melting into ingots.

Before the plant was completed and when the Company needed as much technical advice as it could get World War II started and it was cut off from all direct contact with Austria. While some contact was made through Italy, this also ceased when Italy joined the war on the side of Germany.

The plant struggled on throughout the war, but production runs seldom lasted more than a few days at a time. The problem of converting magnesium dust into ingot metal on a commercial basis was never solved. The foundry was, however, kept busy producing alloys for incendiary bombs, using magnesium from Magnesium Elektron and metal imported from Canada and the U.S.A.

It is worth recording that Kaiser Aluminum Corporation in America worked the Hansgirg process during the war and also found difficulty in the dust-to-metal stage. However, when the military authorities realized that a mixture of magnesium dust and oil made a most effective incendiary material the metal stage was forgotten in favour of the dust-oil mixture rumoured to have been known as Goop.

A bright spot in this otherwise rather drab story was the activities of the plant development and metal research sections where much useful work was done, these sections being largely unaffected by plant production difficulties.

When the war in Europe ended, output of magnesium in Britain, Canada and U.S.A. greatly exceeded demand with little prospect of the latter increasing. The Magnesium Metal Corporation plant at Swansea was closed down and the site with the plant was sold, a number of the staff and workpeople being taken on by one or other of the parent companies.

CHAPTER TWELVE

1939-1945
The War Years

IMPERIAL SMELTING DELIVERS THE GOODS—EFFECTS OF THE
WAR ON BRITISH INDUSTRY IN GENERAL AND THE ZINC SMELTING
INDUSTRY IN PARTICULAR—RECRUITMENT, PRECAUTIONS AND
AIR RAIDS—EFFECTS OF THE WAR ON THE TEMPO OF PRODUCTION
GOVERNMENT CONTROL IN ZINC PRODUCTION AND RAW MATERIALS
FURNACE EXTENSIONS—DILUTION OF LABOUR FORCE
RAPID INCREASE IN DEMAND FOR ZINC ALLOYS AND
EXTENSION OF PRODUCTION TO BLOXWICH WORKS
SUBSEQUENT DECLINE—TECHNICAL WEAKNESSES OVERCOME
SULPHURIC ACID COMES UNDER GOVERNMENT CONTROL
CADMIUM AND AIRCRAFT BEARINGS—ZINC DUST,
ZINC OXIDE AND SMOKE SCREENS—LITHOPONE EXPORTS
ALUMINIUM FLUORIDE AND HYDROFLUORIC ACID COME ON THE
SCENE—CUPRINOL PROSPERS—FERTILIZERS AND THE FOOD EFFORT
THE MODEST PROFITS OF WARTIME

THE PERIOD COVERED by the Second World War has a special significance in the history of Imperial Smelting Corporation. Quite apart from the direct involvement in hostilities through air raid damage and recruiting, which affected most of British industry, the ability of the Company to provide the goods which it had been created—too late—to provide during the First World War, was tested for the first time. It survived the test with conspicuous success.

It is becoming possible now, over twenty years after the event, to make a tentative historical assessment of the effects of the Second World War on the industrial life of the nation. Industry took a further significant step away from the ruins of Victorian free enterprise which had survived the First World War. A condition of chronic unemployment running into two and sometimes three, million out-of-work was exchanged for a condition of over-full employment and rapidly rising wages. Scientific research took another large leap forward, which is one of the few welcome consequences of modern warfare, and Protection in various forms became an accepted part of world commercial relationships.

All these trends affected Imperial Smelting.

The war and old age changed the composition and atmosphere of the Board very rapidly in 1939–40. Sir Cecil Budd retired in March 1939, Oliver Lyttelton (later Lord Chandos) departed in September 1939 to become Controller of Non-Ferrous Metals and, in October 1940, President of the Board of Trade. Lord Horne died in September 1940, Sir Clive Baillieu (later Lord Baillieu), who had succeeded Lord Horne as Chairman on 17 October 1940, left to become Director General of the British Purchasing Commission in the U.S.A. in January 1941. The change continued even more rapidly after the war and the overall result has been the replacement of the pre-1939 Edwardian type of Board of distinguished 'Captains of Industry' of independent means by a 'managerial revolution' type of Board composed entirely of dedicated executives in the wholetime pay of the Company. The wartime Board had features of both eras, with the emphasis on the old rather than the new, while in practice the Company was run by John Govett and L. B. Robinson during these years.

The effect of the rapid change from gross unemployment to over-full employment on the fortunes of the industry, particularly in labour relations and the working of the hand-operated zinc distillation furnaces, will be apparent as this chapter progresses.

Of far less effect on the industry, most regrettably, was the post-war survival of Protection as an accepted feature of international commerce and the continuing history of the British zinc producers' failure to secure adequate national protection in an otherwise highly protected worldwide zinc industry is contained in Chapter 15.

Finally, in the sphere of research, the war undoubtedly stimulated scientific advance in the British zinc industry but cannot be said to have produced sensational results comparable to developments in other industries, such as radar, titanium and jet propulsion.

Admittedly research on the blast furnace process continued but this was a continuation of work on a pre-war idea. It was the fault of the Government, rather than of Imperial Smelting, that the 1939–45 War did not see the rapid development of this process to the commercial stage of production. The circumstances were that the Government again wanted a rapid increase in zinc production and Imperial Smelting's first reaction was to offer to double its modern vertical retort capacity at its own expense. However, as this increased capacity would take over a year to build, the Government insisted that Imperial Smelting should adopt the quicker solution of increasing horizontal retort capacity. The Company were opposed to this facile solution as they foresaw that, in conditions of overfull employment, it would rapidly become impossible to find suitable labour to operate the archaic horizontal process but gave way to the Government's insistence on condition that the Government finance the building of these extra horizontal furnaces.

The details of these negotiations and of how the Company were eventually proved right on the labour question are given later in the chapter. The zinc industry's research effort in the Second World War, however, was confined by this decision, and by the general wartime drain on scientific manpower, to more mundane paths than development of a revolutionary smelting process. These included improvements in working conditions and production techniques, cadmium alloy bearings for aircraft, increasing the amount of cadmium recovered during smelting, a considerable amount of technical service work in connection with Mazak die castings for the armed forces, development of rapid spectrographic and polarographic methods for the analysis of high purity zinc and Mazak, processes for the extraction of magnesium oxide from sea water as a possible raw material for magnesium metal production, and other unspectacular but solidly useful activities.

Of the extent of the direct involvement of Imperial Smelting in the war, however, there can be no doubt.

In spite of the opportunity given to workers in certain sectors of heavy industry to avoid conscription by staying at their jobs, no less than 765 out of a total number of employees not exceeding 3,000 in 1939 joined the Forces. Out of these thirty-six did not return and their names have been bequeathed to history in the Company's Roll of Honour.

All the larger works suffered from air raids, the proximity of Avonmouth and Swansea to important ports on the Western seaboard ensuring that they had a major share of the enemy's attention.

Prior efforts had been made to camouflage Avonmouth Works. A letter from the Air Ministry dated 11 April 1939, mentions a flight carried out over the works by the Camouflage Section and sets out detailed recommendations to be carried into effect with as little delay as possible. A certain amount of camouflaging was done with such manpower as was available but everyone realized that the works and Avonmouth Docks were pinpointed by the confluence of the River Avon and the Bristol Channel, and the main protection consisted of mobile smoke generators, deployed according to wind conditions on the roads surrounding the works.

Air raid precautions for personnel were also taken several months before the outbreak of war. By February 1939 provision had been made 'for the protection of 420 employees in the Pottery Building' at Avonmouth and in April 1939 'in accordance with Air Ministry recommendations' far more positive steps were taken including provision of shelters for 1,050 employees at Avonmouth and further expenditure at Swansea. By the end of the Battle of Britain month of September 1940, the total amount sanctioned by the Board for air raid precautions, including training, blackout, alarms and other necessary equipment of those dramatic days, stood at over £55,000.

Interruptions to operations at Avonmouth by air raid alarms were reported from May 1940 onwards. By August, interruptions at Swansea were serious enough to upset furnacemen's earnings which were based on piecework. In the same month there are reports of the shaking of the foundations by falling bombs causing leaks in the vertical retort shafts at Avonmouth. By this time also lithopone production at Orr's Zinc White a few miles from Liverpool was being affected by 'frequent air raids'.

The first recorded damage was to the water mains and doors of a building in the acid plant at Avonmouth on the night of 24 September 1940. On the night of 25 November 1940 a heavier raid damaged the Cuprinol building, the refluxer house, the metal store and other stores and mess rooms in the eastern half of Avonmouth Works. The raids grew more intense and on the night of 16/17 March damage to an important filter press put the cadmium plant out of action for four weeks and reduced acid production by some fifty tons a day.

The biggest raid of all was on the night of 3/4 April 1941, when damage assessed at over £130,000 was caused at Avonmouth. The aluminium sulphate plant was burnt down and was not rebuilt until after the war and the vertical retort plant was put out of action for a fortnight owing to a direct hit on the filter house and breaks in the water mains.

A notable 'incident' arising out of this raid was the dropping of an outsize high explosive bomb on the rock phosphate store of National Fertilizers. Fortunately it did not explode and was removed safely by the Bomb Disposal Authorities after a Company cranesman had shown exceptional bravery in moving his overhead crane directly over the bomb, shifting several hundreds of tons of superphosphate, and lifting the bomb on to the Bomb Disposal lorry. There was a typically British anti-climax to this drama when the Bomb Disposal squad proceeded to park the lorry and the unexploded bomb outside the Company's main office block and go into an adjoining shed for a cup of tea.

The following morning the German Radio broadcast a statement that the Luftwaffe had attacked industrial targets west of Bristol causing heavy damage.

Considering, however, that a total of 130 high explosive bombs, in addition to many hundreds of incendiary bombs, fell on Avonmouth Works in these months up to early June 1941, when raids on the Bristol area dwindled away to nothing, the damage done was in reality mercifully light and only one employee was killed.

None of the other works were damaged except the comparatively small works at Seaton Carew on the outskirts of Newcastle where a lone daylight raider, many months later, on 6 January 1942, caused some £15,000 worth of damage by blowing up part of the pottery. Swansea Works was particularly fortunate to escape damage in the three consecutive raids in February 1941 which completely destroyed the centre of the town.

This period of air raids can be seen now as a turning point in the history of the efforts made by the zinc smelting industry during the war to provide goods required by the nation.

After the numerous troubles and breakdowns incurred in starting up a variety of new plants and taking over a variety of old companies in the early 'thirties Imperial Smelting can be said to have overcome most of its production troubles by the time the Second World War broke out. Production was at last on the up-grade, although it had by no means reached the standard hoped for, particularly in zinc production by the vertical retort process.

The first two years of the war removed the commercial obstacle of free competition which might have impeded the industry's progress if peace had continued. Under the protection of Government control for most of them production of all the Company's products rose sharply towards the existing limits of production capacity and, in some cases, beyond, as will be recounted later. Then came the air raid period from mid-1940 to mid-1941. The interruptions and dislocations caused then, together with the greatly increasing drain on manpower as the nation started building up its armed forces to World War dimensions after the shock of Dunkirk, caused a significant decline in efficiency and output in almost all products. In the later part of the war also, raw material shortages, particularly of zinc sulphide concentrates, exaggerated the decline in production in several inter-connected sectors.

When war broke out, on 3 September 1939, zinc production had received only three months previously, on 26 May, the stimulus that Imperial Smelting's management had fought for ever since 1932—an increase in the import duty on zinc from 12/6 to 30/– and the promise of a subvention payment of 10/– a ton on every ton of Empire zinc imported and sold in Britain, provided that Imperial Smelting kept its total production below 60,000 tons.

On the outbreak of war dealings on the London Metal Exchange ceased and the new tariff arrangement became partially inoperative before the encouragement which it was intended to provide had had a chance to increase production at Swansea and Avonmouth from the levels to which it had been reduced in 1938. Imperial Smelting, therefore, went into the war with only six horizontal furnaces, recently rebuilt, in operation at Swansea and only two in operation at Avonmouth, together with sixteen vertical retorts and a scheme in hand to increase the total output of the vertical retort plant from 60 to 80 tons per day (29,000 tons per year). Total production in the year ended 30 June 1939 had amounted to only 47,272 tons.

As soon as war broke out the Government approached the Company with the request to increase zinc production. W. S. Robinson reported in writing to the Board on 1 October 1939 that 'discussions with the Ministry of Supply on the contract for acquisition of our zinc output have been almost continuous over the whole month' and that agreement was near. The principal point in this agreement was the undertaking by the Government to purchase the whole of the Company's output of zinc 'estimated to be 60,000 tons per annum at £16 a ton for ordinary g.o.b. ex works plus agreed premiums for higher grades'. The Company would also receive the subvention of 10/– a ton on all zinc purchased by the Government from the Empire producers. The Government undertook to supply the necessary raw material for this production, estimated at 150,000 tons of zinc concentrates per annum and, for this purpose, took over at agreed prices most of the concentrates stocks then technically in possession of the Company whether at its British works, in ships at sea, or in Australia.

Surprisingly enough no difficulty appears to have been encountered in shipping concentrates all the way from Australia to Britain during the first three years of the war. Only one ship, the ss. *Trevanion* carrying 7,872 tons, was reported lost, at the end of 1939, and concentrates were also being shipped from Canada and Burma. The peak was reached at the end of April 1942 when total available supplies amounted to nearly 260,000 tons with 183,742 tons in stock in Britain but only just over 8,000 tons afloat. After that the balance at first gradually and then very rapidly went the other way. By May 1943 stocks in Britain were down to 96,000 tons, by December 1943 to 59,000 tons and by April 1945 to only 37,670 tons. This very drastic drop in raw material supplies was, of course, not sought by the industry and was caused mainly by major changes in the Government shipping programme of which prior warning was given to the Board as early as March 1942. The reasons for these changes were not stated in print. Presumably, however, they must have been connected with the worldwide spread of war operational areas to be serviced by shipping in 1942, particularly for the beginning of the North Africa campaign, and heavy shipping losses inflicted by the enemy in the period before ship replacements from the U.S.A. had had time to build up. By 1944 also another reason for decline in concentrates supplies was apparent at the producing end. A memorandum of unknown authorship, presented to the Board on 18 May 1944, mentions the serious inroads made by heavy wartime demand into the ore reserves of operating lead and zinc mines and the impossibility of extending these reserves by exploration and development owing to the manpower shortage in Australia.

This rise and fall of stocks of zinc sulphide concentrates dominated the production pattern of the Company's main products and most of its ancillary products during the war years to as great an extent, probably, as the air raid period and the increasing labour shortage already mentioned. Subsidiary factors in the last two years of the war were restriction on supplies of coal (at one point Avonmouth Works was down to two week's supply) and of electricity.

In the first two years production was expanded as rapidly as possible in an atmosphere complicated by uncertainties about delivery of plant parts, sudden departures of key employees to the forces and the hygiene, safety and operating difficulties caused by trying to 'black-out' the roasting and smelting plants.

In December 1939, after the discussions with the Government's Controller of Non-Ferrous Metals outlined earlier in this chapter, the Company agreed to install two more furnaces at Avonmouth to produce 8,300 tons more a year and four more furnaces at Swansea, two of which would work on zinc ashes, to produce 12,200 tons more a year, i.e. a total increase of 20,500 tons. The Controller agreed in February 1940 that the Government would finance these extensions at an estimated cost of £27,900. In June 1940 further extensions were approved by the Controller at a total estimated cost of £347,200. These consisted of the recommissioning and reconstruction of available furnaces at Avonmouth, Swansea, Seaton Carew and Bloxwich with annual capacity of 28,980 tons and additional roasting and acid capacity equivalent to metal output of 20,000 tons per annum.

The principle on which agreement with the Ministry proceeded in respect of these and other extensions was that the Ministry would pay the cost but that the new plant would belong to the Company. After the war the Company was to pay a fair part of the cost of any plant that it wished to retain and the remainder was to be maintained by the Ministry for ten years and then scrapped, if not still required. The total cost to the Ministry by the end of the war was over £600,000. Most of this was wasted as only part of this extra plant could be used during the war owing to labour shortage and very little of it was of any use to the Company after the war, with the principal exception of the Swansea acid plant extension for which payment was duly made in succeeding years.

The production potential on completion of this extension and reconstruction was 49,480 tons a year from the horizontal retorts, together with 29,000 tons a year from the recently enlarged vertical retorts, but actual production never reached this total potential of 78,480 tons and the following tabulation best shows the fluctuations in production caused by the factors outlined above in the war years:

	1939	1940	1941	1942	1943	1944	1945
Vertical Retorts	20,496	23,036	21,350	20,711	21,262	23,828	25,078
Horizontal							
Distillation	29,580	37,044	42,667	44,919	40,099	40,311	35,925
Bloxwich	891	1,841	2,212	2,321	2,294	2,372	370
Seaton Carew	—	—	2,450	5,050	6,037	6,247	1,081
Total	50,967	61,921	68,679	73,001	69,692	72,758	62,454

This table shows up the effect of the air raid period of 1941–42 on the vertical retorts and the effect of shortage of concentrates and skilled labour on production as a whole after the peak year of 1942.

The effect on costs of dilution of the labour force with unskilled labour was increasingly drawn to the Government's attention as the war progressed and, in September 1942, the Government had to agree to a reduction in the number of horizontal distillation furnaces in operation at Avonmouth from 5½ to 4 and from 12 to 10 at Swansea to counteract this dilution of skill by means of shrinkage of the labour force.

It will be noted from the table above that zinc smelting operations were restarted on two furnaces at Seaton Carew in April 1941, as part of the subsidized plan agreed with the Government and continued, with the number of furnaces in operation gradually increasing to five, until the end of March 1945 when smelting operations closed down again on the instructions of the Ministry.

At the Delaville Works at Bloxwich one furnace had already been brought back into operation to produce zinc from zinc ashes by the time war broke out and operations were extended up to three furnaces for a brief period as the war proceeded. These operations came to an end in February 1945, again on instructions from the Ministry.

Part of the zinc produced by the vertical retorts was, of course, passed to the refluxer for upgrading for Mazak production at Avonmouth. In October 1939, the refluxer produced just under 900 tons of high-purity zinc but for the first few months of the war there was a decline in demand for zinc alloys for die casting owing to a falling off in demand for non-military uses. However, the Ministry of Supply was quick to realize the potential importance of die casting parts to its ammunition requirements and the appointment of one of the Company's own alloy experts as Deputy Controller of Non-Ferrous Metals with particular responsibility for the die casting industry, had a stimulating effect on promoting extension of the use of zinc alloys from early 1940 onwards. Preference was given to direct alloying as the more rapid method and by May 1940 both the eighteen tons a day refluxer and the newer thirty ton refluxer were in constant use to produce high purity zinc for this purpose. In spite of the general shortage, supplies of the aluminium additive were released without delay by Aluminium Control owing to the growing demand for zinc alloys for ammunition production.

To meet this demand Avonmouth's existing capacity was eventually boosted by various devices to some 600 tons a week but this was the maximum possible without further plant expansion.

Accordingly, the Ministry of Supply, who were not anxious to concentrate all zinc alloy production plant in one vulnerable area during the period when air raids were increasing in intensity, agreed with Imperial Smelting that the Government should finance erection of a further plant to produce another 280 tons of zinc alloy a week on the Delaville Spelter Company site at Bloxwich instead of at Avonmouth.

As all the high purity zinc produced by Imperial Smelting was already being used in direct alloying at Avonmouth the Controller of Metals had to arrange to import Four-Nine zinc produced by the electrolytic zinc plants in Canada as feed for the new Bloxwich plant. This opened the way to a significant change of policy comparable to the admission of the Trojan Horse into Troy, as Canadian zinc remained the basic feed at Bloxwich in the post-war years when Imperial Smelting shut down its high purity zinc production and its alloying operations at Avonmouth.

The final stage of expansion came when, as a safeguard against the refluxer plant breaking down at some future time, it was decided to install at Avonmouth an Ajax Wyatt melting furnace of 280 tons a week capacity to produce alloy from solid metal if required. This started up in June 1941, and rapidly superseded direct alloying as it was found capable of producing up to 700 tons a week.

Output at Bloxwich also increased rapidly after the introduction of incentive schemes for those working on the plant so that, by the end of 1942, Imperial Smelting's total production capacity for zinc alloy was about 1,250 tons a week.

However, by this time a decline in demand from the Armaments Division was already affecting zinc alloy production and a second Ajax Wyatt furnace, which the Ministry had asked Bloxwich to install, was completed but never used.

The reason officially stated to the Board was that this decline was due to the general zinc position but the more likely reason was that, by the end of 1942, the Government found themselves overstocked with ammunition made from zinc die castings which had only a limited application in armaments at that time (e.g. for fuses, trench mortars, bombs and grenades) but very short of ammunition requiring brass, which is usable in a wide variety of ammunition components.

Thus, in July 1943 production of Mazak had to be reduced to 350 tons a week in conformity with the reduced Ministry of Supply programme and at the end of that year production of Mazak at Avonmouth finally ceased, capacity at Bloxwich being adequate to meet the reduced demand, although even there production was intermittent for the rest of the war.

In spite of these changed circumstances, however, Imperial Smelting continued, as in the early years of the war, to supply the whole of the British die casting industry's requirements of zinc alloy even though, for certain non-military uses, it had to eke out supplies of Four-Nine zinc by the production of a brand called 'Remaz' to British Standard 1141 from remelted redundant castings*. This began in April 1943 and went on for the rest of the war. Although dictated by the exigencies of the war, this was, of course, a step in the wrong

---

*From 1940 also limited batches of a brand called 'Durak' were produced from time to time in an endeavour to meet demand from civilian customers. This was based on the Mazak 5 formula but contained secondary instead of virgin aluminium; otherwise, unlike Remaz, it maintained the specified limits for known deleterious impurities.

direction, permitting, as it did, impurity contents higher than those in BS 1004. Soon after the end of the war BS 1141 was withdrawn and production of Remaz ceased.

Great emphasis has been placed in this chapter on the break-through made in Mazak quality and quantity during the 1939–45 War.

Under the high pressure of war demand and danger inseparable from air raids it was inevitable that there would also be production troubles in this relatively new field of zinc alloy die casting. For example, an extract from the Company's files reads 'We have made a careful examination of the position further to the complaint made in respect of one of the zinc-manganese rolling slabs having contained a spanner embedded in the casting'. More generally, in the early days of the war a scare arose from the phenomenon known as low temperature embrittlement. At sub-zero temperatures Mazak (as do some other materials) suffers a considerable loss in impact strength. No other physical property is affected and the drop in impact strength is restored as soon as temperatures above zero are regained. Since the war was being fought in areas ranging from the Tropics to the Arctic it was essential to ensure that this weakness of Mazak should not cause failures. Together with the Armaments Research Department at Woolwich an extensive series of tests was carried out. Principles of design which allowed an adequate margin of safety were agreed and, to provide an additional margin, the copper-containing alloy, Mazak 5, was specified for stressed parts. Due to the presence of copper the onset of cold embrittlement occurred at a lower temperature than with Mazak 3. As a result of this work the use of Mazak for almost all previous applications and many new ones was permitted and there was no reported case of failure from cold embrittlement throughout the war.

It was in this period also that the widespread use of die cast zinc alloy for armament parts, particularly fuse components, and the possible catastrophic result of failure in store or in service, led to the drafting and adoption of British Standards 1003 and 1004 for high purity zinc and zinc alloys for die casting. These went a long way in maintaining the necessary high quality and remained as basic grades after the war.

The other product in which Imperial Smelting made a major contribution to the war effort was, inevitably, sulphuric acid. By October 1939, when sulphuric acid production and consumption came under Government control, the large contact plants at Avonmouth and Swansea were working to capacity and 280–300 tons a week was being produced from imported sulphur at Seaton

Carew. By January 1940 the chamber plants at Newport and Pontardawe had also been brought back into production and total production from the five acid producing works was about 19,000 tons a month. To the continuing demands from peace-time users in the steel, fertilizer, cellophane and other industries were added orders for 96 per cent strength acid from munitions factories, beginning with the Royal Ordnance Factory, Pembrey. No. 6 acid unit at Swansea was extended in the first half of 1940 and 20,000 tons a year extra capacity added also to the Avonmouth plant. By late 1940 deliveries of 96 per cent acid were also being made to the I.C.I. alkali plant at Hayle and soon to the munitions factories at Holton Heath, Waltham Abbey and Wrexham.

By early 1941, however, demands from munitions factories, for various possible but unknown reasons, became more erratic and the Controller allowed non-munition users, such as National Fertilizers, to take more. On 1 April 1942 control was tightened still further when the Controller took over complete control of output and distribution, the acid industry becoming in effect agent for the Minister. The first result for Imperial Smelting was an order closing Newport and Pontardawe Works in order to reduce stocks. By April 1943, however, demand was already starting to rise again and when, in September 1943, deliveries rose to over 23,000 tons in one month the Controller ordered Newport to be brought back into production using spent oxide from gasworks as raw material. This had already been used, instead of sulphur, at Seaton Carew from the end of 1942 after encouragement from the Controller, which reveals that imports of sulphur were becoming more difficult. Nevertheless sulphur was still used extensively at Avonmouth and Swansea during the last two years of the war to maintain the level of acid production after the rapid decline began in supplies of zinc sulphide concentrates.

Imperial Smelting's other products also played a useful, if less important, part in the war effort, and those which were by-products of zinc also suffered in similar fashion from the uncertainty surrounding supplies of raw materials.

The cadmium plant at Avonmouth was extended in the first half of 1940. Output rose from about six tons to about eighteen tons a month and much of this went into the production of cadmium nickel alloy mainly for use by the aircraft industry for Bristol Beaufighter engine bearings. Cadmium came under the control of the Controller of Non-Ferrous Metals from 1 January 1942 and the price was fixed at the level of 5/6 a lb.

Imports of zinc dust were severely restricted by the war so that the Company's production effort increased rapidly from only nine units at Bloxwich in September 1939 to twenty-eight at the end of June 1940. Fricker's also made zinc dust from time to time in small quantities of 50–100 tons a month as compared with 300–400 tons a month of Delaville zinc dust made at Bloxwich. All Fricker's output of zinc dust at this time was being exported for use in the gold mines in South Africa until shipping difficulties led to the decision that from 1 February 1942 all Delaville and Fricker's zinc dust should be sold in the home market at a fixed price of £35. 10s. 0d. a ton. Later in 1942 the Chemical Defence Department wanted zinc dust for use in producing smoke screens and the Company agreed to install a plant to produce thirty tons a day of this by an air-blown process at the Government's expense. This plant came into operation in April 1943 but was shut down by the Government in April 1944. It was also agreed to install a plant to produce chlorosulphonic acid mixture at Avonmouth, again for supply to the Government for smoke-producing purposes and using among other things 300 tons of sulphuric acid a week. This started up in April 1944 and also ran for less than a year. Smoke requirements appear to have increased as the war proceeded as, in July 1943, it was decided that the entire output of Fricker's zinc dust should go to the Government for this purpose leaving only Delaville zinc dust for the home trade. This Government demand, however, lasted only until September 1944 when the Fricker's zinc dust plant was shut down. Direct grade zinc oxide produced by Fricker's was also pressed into service for use in manufacturing smoke bombs for a period in the middle of the war.

Fricker's indirect grade zinc oxide started the war in tremendous demand, as a pigment much used in camouflage work, and demand exceeded supply for quite a long period in 1941. However, production had to be curtailed when supplies of slab zinc for this purpose were reduced by the Controller after the first two years of the war. Some of the retorts were converted instead to zinc dust production.

Lithopone also was in great demand as a pigment early in the war and all four units at Widnes were in production for most of the period. Exports were prohibited when war broke out and the Controller of Non-Ferrous Metals imposed a maximum price of £16. 15s. 0d. a ton in October 1939. (This was increased several times during the war.) W. S. Robinson was particularly concerned, however, that the export market which Orr's had built up in Canada should at least be kept alive during the war period and in November 1939 the Ministry of Supply permitted small quantities to be exported to Canada and Australia. Approval was given in April 1940 for an increase in production capacity at Widnes from 37,000 to 44,000 tons and a surprising feature of these war years is that exports of lithopone continued to expand even after lithopone sales came under much stricter control at the end of 1941.

Exports were permitted to such an extent that W. S. Robinson's original plans to get round the export difficulty by erecting production plants in Canada and Australia appear to have been dropped as the months went on.

Aluminium sulphate production at Avonmouth, although booming after the cessation of foreign imports at the beginning of the war, ceased entirely after the plant was destroyed in the air raid of 4 April 1941. Building of a new plant was not discussed until 1945. Meanwhile, production of the new chemical products, hydrofluoric acid and aluminium fluoride, started in 1939, the main outlet being British Aluminium.

The first aluminium fluoride plant, based on the design of Herr Schuch of Hanover, was an almost catastrophic failure and after three months of painful effort it was decided that a second plant should be built based on the designs suggested by the previously discarded original work. There was no time to carry out pilot plant trials and a plant was rushed up which was completely mechanized. An agitated vessel was installed to carry out the initial reaction, followed by a centrifuge to separate and partially dry the aluminium fluoride. The greatest risk was taken by the installation of a 30 foot long kiln internally heated by coal gas, particular care being taken to control combustion conditions accurately. This was, of course, the biggest step into the unknown.

The plant worked well almost from the start. Few teething troubles were experienced and an output of 30 tons a week was soon reached and maintained steadily throughout the war, satisfying in full the demands of British Aluminium.

Towards the end of the war it was realized that an even better method of producing aluminium fluoride would be to carry out both the reaction and the drying stages in one vessel using the newly developed fluidized bed techniques. A pilot unit was built in which partially dried aluminium oxide was fed continuously into a heated vessel. Hydrofluoric acid vapour was fed into the bottom of the vessel through a perforated plate at a rate which kept the alumina in suspension and allowed the reaction and drying stages to occur together almost instantaneously. Dried aluminium fluoride was removed continuously from the top of the vessel.

At the end of the war the aluminium fluoride plant was shut down having served its purpose. British Aluminium could obtain the product it required from the very much larger and, therefore, cheaper plants built to serve the aluminium industry in Canada. The work on the fluidized unit was so promising, however, that an arrangement was made with Ugine in France whereby they would build a full-scale production unit to the same design but give Imperial Smelting full operating know-how should they wish to use it. This was done and a unit was built to produce twelve tons a day of fluoride. This has worked very satisfactorily and a second and similar unit has recently been constructed.

Thus, although aluminium fluoride production at Avonmouth was largely a wartime effort and did not continue, it served its purpose to the full. It also provided experience of hydrofluoric acid production which led directly to post-war expansion into the fluorine field. This will be described in the next chapter.

Cuprinol flourished in the war more than it had ever done in the pre-war years and was used widely for rot-proofing tent canvas, sandbags, timber, equipment for use in tropical conditions in the Far Eastern War, and numerous other uses.

The Newport and Avonmouth fertilizer plants and the slag grinding plants of National Fertilizers continued in full operation throughout the war and continued to be managed by Imperial Smelting even after National Fertilizers merged with Fisons. Production was at times affected by air raids and shortage of sulphuric acid but the indirect contribution made by this Company to aiding intensive cultivation of food during the war formed an important, if mundane, part of the war effort.

After the country's experience of war profiteering during the First World War 'profits' was a word seldom mentionable in the second war. Wartime taxation of profits and Government levies peculiar to various industries were heavy but, in spite of this, Imperial Smelting was able, at last, to make a steady, if modest, financial profit and declare a dividend of 4 per cent on its Ordinary Shares each year. This was mainly due to the contract with the Government which assured a zinc price at a level at which a small profit could be made and provided that the Government would pay the increased costs arising during the war. It is difficult to disentangle the plain facts of profitability from the wartime accounts, which are complicated by numerous payments to and from the Government, but it seems certain that, in the absence of competition and with the protection of a controlled price for some products, most activities were showing a real profit, with the probable exception of zinc production from the vertical retort plant.

The hard problems of running the business to make a real profit in peace-time conditions returned, of course, as soon as the war effort ceased and rapidly became harder as wartime controls and post-war shortages disappeared after 1952.

CHAPTER THIRTEEN

1945-1967
Post-war Boom
and Slump

THE DIFFICULTY OF WRITING 'CONTEMPORARY HISTORY'
THE INCREASING COST AND RISK OF NEW METALLURGICAL
PROCESSES IN THE POST-WAR WORLD—GENERAL POST-WAR
SHORTAGES, CONTROLS AND THE KOREAN WAR PRODUCE
PROFITS FROM A CONSERVATIVE POLICY UNTIL 1956
VAST CHANGES SINCE 1957—THE POST-WAR RECONSTRUCTION
'PROGRAMME' AND MODERNIZATION AFFECTED BY UNCERTAINTY
OVER THE FUTURE FOR THE IMPERIAL SMELTING FURNACE
A POLICY ON MAZAK PRODUCTION FROM IMPORTED ZINC
SLOWLY EMERGES—NEW METAL RECOVERY VENTURES
THE WILKINS-POLAND ELECTRIC FURNACE—THE WAELZ KILN
PROJECT MISSES THE BUS—TITANIUM DISAPPOINTS
BERYLLIUM AND THE ATOMIC FUTURE—SULPHURIC ACID
BOOMS AND DECLINES—LITHOPONE GRADUALLY WITHERS AWAY
DEVELOPMENT OF BARIUM CHEMICALS—ENTERPRISE AND
EXPANSION IN THE WORLD OF FLUORINE—ISCEON
OTHER PRODUCTS, OPPORTUNISM AND INSPIRATION

IT IS DIFFICULT to write a completely objective account of the past twenty years of Imperial Smelting Corporation and of zinc smelting in Britain for the obvious reasons that make any writing of 'Contemporary History' a hazardous task. Many of the personalities involved are still living and still in possession of strongly held personal views on various aspects of policy. Obviously also, and in accordance with accepted commercial practice, many Company secrets must remain unpublished for the present, except where they are covered by internationally recognized patents. Above all, no branch of knowledge is more subject to sudden and startling new discoveries and changes of outlook than the physical sciences. It would, therefore, be unwise at this early stage to claim permanent importance for some of the major developments pioneered since the 1939–45 War by the country's sole zinc producer, Imperial Smelting Corporation, when they may prove in the long run to be passing phases. The glamour of the revolutionary nineteenth-century invention of the steam locomotive moving on fixed tracks was soon eclipsed by the invention of the motor car, and the aeroplane soon ousted the airship. In both cases a completely different 'process' and not a development of the previous 'process' was involved.

It may well be that the development since the war of the Imperial Smelting process for the simultaneous production of zinc and lead from a blast furnace will be one of the greatest discoveries ever made by the zinc industry. The process has, since 1957, been adopted by over a dozen zinc producers overseas but a great deal more research work would have to be done before it could fulfil a dream worthy of the medieval alchemists—a cauldron into which is flung a mixed ore and from which emerges in pure or nearly pure form different streams of high and low grade zinc, lead, copper, gold, silver and all the metallic constituents of the ore transmuted into metal. Such a dream may never be fulfilled or may be frustrated by the prior invention of a cheaper and more efficacious process. The answer to this may be known by the end of the century, provided, of course, that zinc, lead and the principal non-ferrous metals of the present day have not by then been superseded by more recently discovered metals or synthetic materials such as plastics. No one can prophesy at this stage. For this reason involvement in the metal producing industries of the country must be a perpetual and increasingly expensive gamble —a gamble, nevertheless, which is vital to the continuing prosperity of the community.

In the case of the Imperial Smelting process this gamble has already cost the Company which invented it over £4 million in research costs in addition to another £4 million spent on its own full-scale Imperial Smelting furnace and ancillary plant at Swansea and £14 million on the new complex at Avonmouth.

This is a measure of the financial vulnerability of metal producing industries generally to sudden changes of process and metal using fashion. It is one explanation also of the illusion that there was a lack of drive and imagination shown by the Board of Imperial Smelting Corporation in the ten years immediately following the war. The new process had not yet reached the stage at which it could be operated on a profitable commercial scale but there was a certain measure of faith that it would be a 'winner'. In these circumstances the only alternative was to continue operating the industry's existing zinc producing processes on which, ultimately, most of its chemical sidelines also depended and not spend too much money on them or on other bright ideas of the powerful older men or the rising young men in the industry.

This chapter gives an account of this mundane but necessary task. The next gives an account of the much more spectacular achievement which forms the background to the years since the 1939–45 War.

Fortunately there was every inducement for letting the pre-war shape of operations continue in the general economic circumstances in Britain during the years up to 1956–57. Added to a general post-war shortage of all the industry's products, price and distribution controls by the Government continued until the end of 1952 and early 1953 for the industry's main products—zinc, sulphuric acid, and lithopone. This assured to the industry markets and, at least, reasonably profitable prices for all its output during these years. It did more than this in the eighteen months after June 1950 when the Korean War broke out and sent the Government controlled price of zinc up from £95. 10s. 0d. in April 1950 to a peak of £190 on 14 June 1951, while the free market price went as high as £200–£205 on 18 February 1952.

The return in October 1951 of a Conservative Government bent on scrapping continuing wartime controls wherever possible and the ending of the Korean War in July 1953 brought, of course, some months of worry to the industry when the zinc price, after the reopening of the London Metal Exchange on 1 January 1953, dropped to as little as £63. 15s. 0d. a ton on 23 April 1953 (i.e. a drop of over £125 in the effective price to the industry in less than eighteen months). However, the post-control boom of the 'never had it so good' society soon redressed the situation and by April 1954 the London Metal Exchange zinc price again exceeded £80 and by January 1955 was over £90. The price continued in the region of £100 until 1957 and most other products of the industry remained at a corresponding level of prosperity in that period. Since then the industry has been subject to an increasing tempo of boom and slump to be outlined in the last chapter of this book.

Imperial Smelting, of course, made steady profits in these immediate post-war years but the feeling still lingers on that the industry might have profited a great deal more from the 'sellers market' of the times both financially and by grasping the opportunity afforded by easy times to re-equip and reorganize itself to meet the possibility of a post-war slump which undoubtedly lay somewhere ahead.

As events happened, when J. R. Govett died on 28 November 1956 shortly after relinquishing the post of Chairman which he had held since early 1941, the processes and works actually operating in the Imperial Smelting Group of companies were almost identical with those which had been operating at the outbreak of war in 1939. The only exceptions, apart from a few, at that time, small tonnage chemical additions such as fluorine and barium chemicals, were that aluminium fluoride production had ceased and that the Waelz kiln plant for the treatment of zinc furnace residues and two small zinc blast furnaces had started up. In the succeeding ten years to the present time, under the policy carried through by L. B. Robinson, M. I. Freeman, and their successors, the face and operations of Avonmouth, Swansea and Widnes, which were the three largest works in 1956, have changed almost beyond recognition, two works have been disposed of, and several subsidiary works acquired.

The contrast between these two periods must not be taken to imply that in the ten post-war years the management were content to sit idly by and let the industrial scarcities of the time reap the profits for them. There was a wealth of ideas on new projects and new ideas on old processes which were being produced by numerous individuals, departments, and meetings in those years. It would be totally untrue to say that all this effort achieved nothing. The fundamental ideas, on which most of the really worthwhile commercial developments of the past ten years were based, were worked out in that period.

However, for all the ideas, the admixture of new blood at the top after the war, the opening up of wider financial horizons through the final merger in 1949 with The Zinc Corporation of Australia in 'The Consolidated Zinc Corporation', and the undoubted boom in demand for all Imperial Smelting's products from 1945 to 1952, the Board cannot be said to have pursued a particularly adventurous policy. The Chairman's published statement for the accounting year ended 30 June 1947 admitted this:

The net profits for the year ended 30th June last showed a substantial increase. Two factors are in the main responsible for the marked improvement, viz: a wider margin between costs of raw material and operation charges on the one hand and realised price on the other, and the effects of the rigidly conservative policy followed by the Board during the past few years. It would be unwise to anticipate a continuance of the former.

The net annual profits continued to improve by substantial amounts right up to the end of 1955, with a brief decline in 1952–53 resulting from the disturbance caused by the ending of Government control of the zinc price. This brief decline, however, was sufficient to induce the Board to impose late in 1952 an embargo on further capital expenditure except in very special circumstances. The cash resources of the Company had at that time been depleted by the limited post-war reconstruction programme but, even when they recovered rapidly from around £700,000 to over £2 million again, the embargo on capital expenditure was not revoked. This was, perhaps, symptomatic of the cautious outlook of the Board at the time.

The natures and objectives of the post-war reconstruction 'programme' are not easy to piece together at this distance of time as they were nowhere laid down in any traceable document and a great deal of the 'programme' was obviously dictated *ad hoc* by the circumstances of the time. Most of these post-war circumstances were common to British industry generally—shortages of labour and materials with accompanying rising wages and costs and a great deal of Government restriction and new Government imposed costs, such as payments towards the apparatus of the Welfare State. The Labour Government was in office from 1945 until October 1951 and the Korean War, which had broken out in June 1950, lasted until July 1953, and brought with it at one period a revival of Control Orders and shortages in industry similar in some respects to those of the 1939–45 War. For example, Imperial Smelting were again asked by the Government to state what increases they could effect in zinc production capacity. The Board, however, was by this time growing accustomed to the commercial dangers of submitting to this type of wartime Government persuasion. Their prompt reaction was to debate whether they should, before replying, seek assurances from the Government that increased capacity would be protected by guaranteed market margins or in other ways against the menace of dumping of continental zinc as soon as the emergency was over!

It was, in fact, not until the end of the Korean War in 1953 that trading conditions in the zinc industry can be said to have returned to 'normal' for the first time since 1939.

Meanwhile, as is also obvious from other statements after 1945:

The main efforts of the management have been to complete the change-over from war to peace, concentrating on improved plant and process efficiencies, and overhauling the plant, machinery, and property after the severe strain of the war years.

(Statement of late 1946)

The Chairman's statement dated 14 May 1953, however, makes it clear that 'modernization' was also an objective as well as reconstruction—'a considerable part of our modernization scheme, which we instituted as soon as conditions permitted after the war, has already been installed'.

It remains therefore to summarize what was done to restore and what was done to 'modernize' and, beyond that, to give an account of what was done with a view to future development and expansion.

As regards restoration the only total loss to be restored was the aluminium sulphate plant at Avonmouth, destroyed by enemy action in 1941, and this was rebuilt to a modern design and restarted during 1947. It was designed to produce 20,000 tons a year, being the tonnage by which it was estimated that the immediate post-war demand for the home and export trade would exceed the existing production capacity of the country. Machine tools throughout all works were replaced or renovated, medical, dental and optical clinics were provided for both Avonmouth and Swansea Works and the New Passage Hostel at Severn Beach near Avonmouth, which had been acquired during the war, was used to provide temporary housing for labour recruited for Avonmouth from other parts of the country. Extensive air raid damage to housing in Bristol during the war had added to the problems of recruiting labour for that area afterwards.

The pattern of capital expenditure on restoring or improving output of the Company's main products, zinc and sulphuric acid, was dictated by a complex variety of circumstances. The wartime arrangement under which Imperial Smelting's entire production of zinc was sold to the Ministry of Supply at a controlled price, the Government being responsible for supply of concentrates, was replaced from 25 May 1946 until the end of 1952 by a new arrangement. Under this the Company still agreed to sell all its zinc to the Government on an 'evergreen' renewable contract but had to find the necessary raw materials itself which meant that the Company had to borrow an extra £$\frac{1}{2}$ million from the Bank to finance these stocks. There were serious interruptions in the transport of concentrates from Broken Hill owing to rail strikes in Australia, dock strikes in England and Australia and shortage of sea transport for two periods of some months during these years. There were also periodical shortages of suitable labour to work the horizontal furnaces at Avonmouth and Swansea until 1950, which also led to the closing down of part of these furnaces on various occasions as well as the intermittent adoption of a forty-eight hour instead of the twenty-four hour cycle of operation. No one appears to have been unduly worried by any resultant shortage of zinc as the Company was still, in theory, limited to production of 60,000 tons a year under the 'Empire' arrangement of pre-war days and the Canadian companies appeared ready to provide any extra zinc required.

It would also have been an embarrassment to the Company to have been expected to expand production capacity substantially during these years. The more far-seeing Directors did not wish to spend any more money on the horizontal distillation plants because rising labour costs were making them increasingly uneconomic and they had faith that another few years would see the commercial start-up of the Company's new process for producing zinc. Output from the retorts of the vertical retort plant also continued to improve very rapidly after the war with regular production of over five tons per retort per day which was above the original standard set by the inventors, the New Jersey Zinc Company. This improvement and the increase in the number of retorts from sixteen to eighteen more than compensated for reduction of output on the horizontal retort plants but, overall, very little new capital was laid out on either of these zinc producing processes during this decade after the war. There were, however, executives who were campaigning for erecting a battery of vertical retorts at Swansea and 'mechanizing' the horizontal retorts by introduction of various continental methods of charging and stirring out. The Board was probably wise to override them, but the strength of the party who favoured this cautious traditional approach to modernization was considerable in the early 'fifties.

The more pressing zinc problem of this period was the future of zinc alloy production from imported Canadian metal. The actual facts of the years up to the end of the Korean War were that the supply of high purity imported metal through the Government was frequently interrupted. This did not improve the regularity of sales of Mazak alloy although, quite obviously, U.S.A. experience showed that a far larger market awaited development in Britain. The Avonmouth alloy plant never reopened after the war but Board policy on the continuation of production of Crown Special zinc for the Bloxwich alloy plant from the refluxer at Avonmouth underwent several changes of direction in this period until the refluxer was closed down finally as uneconomic in 1956. The question revolved round the proposition that, to be economic, the cost of refluxing Severn metal to Crown Special grade must not exceed the premium paid for imported zinc of similar quality. As the premium was liable to frequent change at the behest of the U.S.A. and Canadian zinc markets and of the British Government and as a considerable amount of capital had been invested in the refluxer plant and in rebuilding it and adding a third refluxer column from 1949 to 1951, the decision on whether to continue operating the refluxer or not was not an easy one to make, particularly when the Korean War introduced an artificial element of shortage.

The eventual solution was a contract negotiated with the Government for purchasing 80 per cent of the Company's requirements of high purity zinc at a discount from overseas sources. This came into effect on 1 June 1956 and brought with it the final closure of the refluxer plant and the end of the affair for a decade. Production of one of the British zinc industry's most profitable and useful products in war and peace, zinc alloy, has thus depended since 1956 entirely on supplies of imported high purity zinc, but the starting up of new refluxers at Avonmouth in 1968 will mean the use of both British and Canadian zinc in British die casting alloy in future.

New metal producing ventures of these years, with the conspicuous exception of the zinc blast furnace, were not so successful.

With the continuous improvement in recovery techniques, what may be termed 'raking over the past' to recover mineral values left by previous generations is a perpetual feature of the world of mining and metallurgy and the satisfactory metal prices prevailing in the early post-war years tempted Imperial Smelting into two abortive metal recovery ventures.

The first and less expensive project was the decision made in September 1948 'To treat low grade brass scrap for the recovery of copper and zinc in metallic form by the use of the "Wilkins-Poland" electric furnace as developed by the Revere Copper and Brass Co. (of New York) Inc.' An exclusive licence to operate the process in the United Kingdom and Australia was granted in March 1949 but the process was never operated by Imperial Smelting, even at Bloxwich where the plant was set up. The plant was reported in January 1951 as nearly complete but a cryptic note adds that 'The possibilities of other uses for the equipment have been examined but none so far is evident'. The fact was that no suitable scrap brass was available for treatment at anything approaching an economic price owing to the outbreak of the Korean War and the plant was eventually sold.

In the same operations report that contains the cryptic note quoted above appears also the first official reference to the second and far more expensive metal recovery project of these years, the Waelz kiln project. In fact this venture eventually proved to be the most costly single project that Imperial Smelting had ever indulged in up to that date.

The Waelz process is a German process for the recovery of zinc and lead oxide from low grade zinciferous minerals and residues and is frequently to be found alongside various metal residue dumps around the world.

This was not the first time that Imperial Smelting had got itself involved with a Waelz project. Installation of Waelz kilns was discussed in 1926 when the Waelz process was being commercialized by Krupp Grusenwerk at Magdeburg and in 1932 the Company had purchased from the Receiver a majority interest in Tindale Zinc Extraction Limited, a small company in remoter Northumberland with plant, machinery and the right to use the residue dumps leased from two noble Lords, the owners of the land. The possibility of moving the plant to Avonmouth was considered and rejected but it was not until October 1937 that Imperial Smelting agreed with the minority shareholders to start up the Tindale Waelz kiln to produce zinc oxide for use at Widnes lithopone works. Some hundreds of tons of zinc oxide were produced in 1938 but after a few months Stanley Robson recommended that the operation should be closed down as uneconomic. The recommendation was accepted and the site was cleared by 1939.

A report prepared in 1949 by Herr Jensen, who had been General Manager of the Unterharzer Works at Oker during the war and had become a Consultant to Imperial Smelting after it, started off the next Waelz attempt. This project was to operate Waelz kilns on the accumulated dumps of horizontal and vertical retort distillation residues, initially at Avonmouth and possibly later at Swansea. It was hoped to purchase the kilns secondhand. The proposal was delayed for two years and the principal reason for bringing it up eventually was the unprecedently high prices of zinc and lead in 1951.

The Board approved the proposal 'in principle' for Avonmouth Works only, at an estimated cost of £650,000 in August 1951. However, as a result of initial delay in receipt of the necessary licences, the prolonged shortage of steel at that time, and prolonged delay in delivery of the kilns, the first of the four proposed kilns did not go into operation until July 1953, the second, a smaller 'agglomeratory' one, a month later and the third in January 1954. By 1953 the zinc price had dropped to around £75 but the kilns were brought into use at first mainly to treat zinc blast furnace arisings as a step towards solving the problem of including these arisings in sinter. Calcined zinc/lead oxides produced by the process were also being used as a suitable addition to the feed of the blast furnace, which was not a use which had been foreseen originally.

Accretion troubles and brickwork failures of the kilns were encountered almost from the beginning and by the time the plant was closed down in 1958 over £900,000 had been spent on it in capital alone. Apart from frequent operational breakdowns the main reasons for closing it were the drop in the zinc price to around £70 in 1957–58 and the availability, by then, of suitable mixed zinc/lead ores and concentrates for the new zinc blast furnaces from overseas. The dumps, of course, remained on the Avonmouth site and the Waelz plant remained there also until September 1961 when it was sold to a dealer. Several voluminous reports were written in the intervening years on the economics of restarting the plant—this is understandable in view of its initial cost—but it did not survive into the era of higher zinc prices after 1961. It is probable that if work had started on this project in 1949 instead of 1951 it would have been at least a moderate success financially instead of a failure.

Two agreements of this period to carry out metals investigation work on behalf of the Government were reminiscent of the earlier magnesium venture and similarly unhappy in their outcome.

The first decade after the war will be remembered, among other things, as a period when extravagant hopes were entertained for the rapid development of certain techniques and materials thrown up in the later stages of the war. The 'peaceful uses of atomic energy' was one of these and the company got involved in this to its advantage, as will be described, over supplies of hydrofluoric acid to the Atomic Energy Authority but rather less to its advantage over beryllium. Titanium was another element much spoken of in these years as a 'wonder metal' and it is no surprise to find it recorded that, in May 1949, expenditure of £10,000 was approved for small-scale research into titanium and zirconium. By November 1949 titanium of 99·4 per cent purity had been made in the laboratory at Avonmouth and consideration was being given to manufacturing 100 lbs. a month for the Government. This intention apparently came to nothing as it is recorded two years later at the end of 1951 that, although the Board were still rather sceptical about the possibilities of this venture, which were not yet fully proved, they agreed to spend £15,000 on a pilot plant. By October 1952, however, the Board had decided that the venture was too risky to continue at the Company's expense but that they would consider carrying on with the work as agents for the Ministry of Materials. In 1954 the Ministry were reported to be interested in acquiring the Company's titanium patents and in March 1955 the Board agreed that the whole position should be reviewed as 'No commercial market has yet developed and the military one is feeble at present'. Above all they thought that the effort being expended would be better spent on zinc smelting activities. Finally, after a visit by the Research Director to study the more developed titanium industry in the U.S.A. in the spring of 1955, the whole project was dropped and the patents lapsed.

A project which had a similar origin as an agreement to do research work on behalf of the Government and which has survived to become a Company project since 1958, was the production of beryllium metal. As early as 1934 it is recorded that a Research Department representative had visited Austria to investigate, among other things, the 'Popper Beryllium Process' but nothing seems to have come of this, although there was much talk of co-operation between Imperial Smelting, Electrolytic Zinc of Australia, and Port Pirie to develop beryllium production within the 'Empire' in these years. Then, in 1949, at the request of the Ministry of Supply, it was agreed to examine at the expense of the Ministry the problems involved in the production of beryllium metal for the Atomic Energy Research Establishment at Harwell by a thermal process involving the reduction of fused beryllium fluoride by molten magnesium. A pilot plant was to be designed and operated to provide data for a production unit to be built later at the Ministry's factory at Milford Haven. In 1951 the Ministry accepted the Company's estimates of the probable cost of beryllium production on their behalf and £32,000 was officially approved for 'beryllium investigation' in 1952. By April 1953 the 'light metal' plant at Avonmouth had been completed but as the whole project was being kept on the secret list at that time there is little recorded information about these early days until 1957 when it was suddenly announced to the shareholders.

The fact was that on 25 May 1957 Imperial Smelting, on behalf of the Government who were financing the project, had started producing beryllium by the thermal process in the new plant at Avonmouth and the ingots produced were being supplied to the Atomic Energy Authority. Another British company, Murex Limited, were at that time also producing beryllium metal by an electrolytic route for testing by the Authority. There were no other producers of beryllium metal in Britain, although I.C.I. and Hawker Siddeley were engaged in its fabrication. The prize dangled vaguely by the Atomic Energy Authority was a big one, namely the prospect of their adopting beryllium for canning the fuel elements for the experimental advanced gas cooler reactor. This method, if successful, would be followed in, possibly, eighteen atomic power stations to be constructed in due course by the Central Electricity Generating Board and this would require the production of about a hundred tons of beryllium metal a year, starting probably in 1963. By 1958 it appeared that the Authority preferred the metal produced by Imperial Smelting's thermal route to metal produced electrolytically and, on this encouragement, the Company bought the small Government owned and Company operated beryllium plant at Avonmouth at the end of 1958. Production was raised to over a ton a year and plans were prepared for erection of an enormous plant to produce the 100 tons a year which the Authority might require. The Consolidated Zinc Board, already pressed to the financial limit by expensive projects for the Swansea zinc furnace, the Sulphide Corporation zinc furnace, and the

Australian aluminium project, then suggested that the venture might best be continued in partnership with one of the two U.S.A. producers who were already firmly established and knowledgeable in this field. Accordingly, agreement was reached with the Beryllium Corporation of Reading, Pennsylvania, in July 1959, and ripened into the formation of Consolidated Beryllium Limited which was incorporated on 21 September 1959. Each partner was to hold 50 per cent of the shares.

The new Company decided very early to extend the scope of its operations beyond beryllium metal and put up a plant for production of beryllium copper master alloy at Avonmouth as Britain had no indigenous source of this product but relied on imports from the U.S.A. and France. Beryllium copper master alloy is used extensively in the production of sparkproof tools and electronic instruments and is a modern alloy in growing demand. A plant of 270 tons a year maximum capacity came into operation at Avonmouth in February 1961 and has been producing master alloy since then, although never yet up to the full capacity of the plant owing mainly to inadequate protection from foreign dumping.

Another venture of the newly formed Company, Consolidated Beryllium, was the acceptance of the offer of the Atomic Energy Authority to sell to the Company their Milford Haven Works, which was taken over in September 1960. This works produced beryllium oxide ceramics in various shapes and for various applications, mainly for the Atomic Weapons Research Establishment at Harwell and certain large electronic concerns.

Two years after these high hopes had been built up the whole Consolidated Beryllium venture was shaken to the foundations by a press announcement by the Atomic Energy Authority in January 1962 that 'The Authority have made satisfactory progress with stainless steel cladding indicating the likelihood of higher fuel temperatures, thinner fuel cans, and longer burn-ups. With regard to beryllium, however, technical difficulties still remain. In consequence the emphasis of development offered for the immediate future will be concentrated on stainless steel and work on beryllium will be conducted on a longer term basis.'

This announcement had the effect of closing the Avonmouth beryllium metal plant and also the beryllium fabrication plants of I.C.I. and Hawker Siddeley. However, hope still remains alive for the future in view of the continuing use of beryllium metal in the U.S.A. in nuclear reactors, missiles, and spaceships and the interest shown by various electronic and aircraft concerns.

Meanwhile Consolidated Beryllium Limited continues as a partnership producing beryllium copper master alloy and ceramics, although the commercial progress of these products in Britain has hitherto been disappointing.

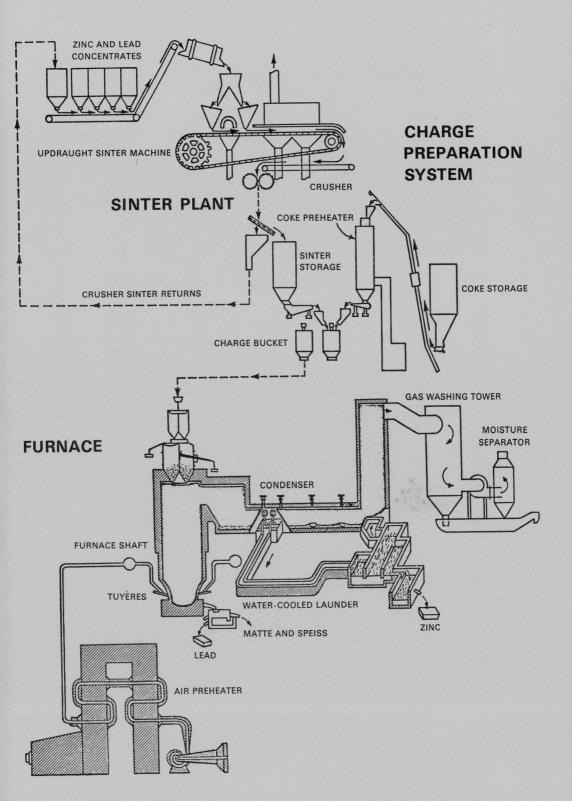

The Imperial Smelting Process for simultaneous production of zinc and lead from a blast furnace.

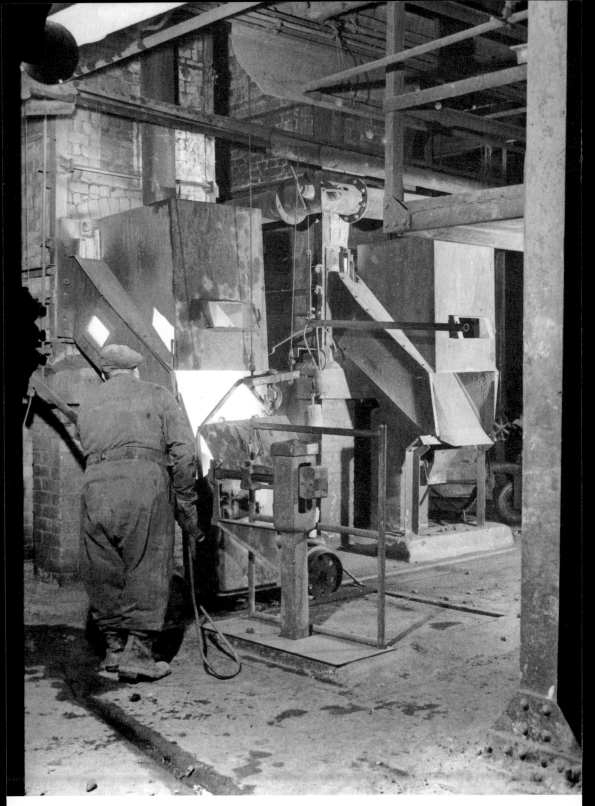

Experimental blast furnace (1947)—Withdrawing hot coke from preheater

Slag running from experimental blast furnace

Another aspect of the immediate post-war years was a boom in the chemical by-products and sidelines of the zinc smelting industry which tempted the management into several new ventures, one or two of which they afterwards had cause to regret.

By contrast with the Government's apparently unconcerned attitude about the peace-time level of home production of zinc the demand for sulphuric acid which, by its nature, cannot be imported so easily, was taken very seriously by the Sulphuric Acid Controller and usage in Britain increased very rapidly after the war. There was a drive to increase fertilizer production in the early years and in the 'fifties the motor car boom and building reconstruction sent up the demand for steel and with it the need for acid for steel pickling. For several years total British demand far outstripped production and total production from all Imperial Smelting plants increased from 157,000 tons in 1945 to an all time record of 264,000 tons in 1954. Obviously, from what has been stated previously, the sulphur content of the limited quantities of zinc concentrates received was inadequate to provide the whole of these record amounts of sulphuric acid and most of the capital spent on 'modernization' in these years was laid out on the adaptation of Avonmouth, Swansea, Newport and Seaton Carew Works to produce the maximum possible amount of sulphuric acid from the sintering process and direct from other sulphur containing materials. As there was a marked shortage of sulphur until new deposits at Lacq in France and in Mexico came into production in the mid-'fifties, the Seaton Carew and Newport chamber plants were operating on spent oxide purchased from gas works at this time having been adapted for this purpose by the installation of three Harris rotary mechanical roasters.

In addition, two rotary Glensfalls sulphur burners were added at Avonmouth, two of these were later installed at Swansea and the existing Avonmouth and Swansea sinter and acid plants were modernized in 1951–52 at a total cost of over £2½ million which was later estimated to be capable of producing an extra profit of £328,000 a year. This included provision of Herreschoff furnaces for pyrites burning in periods of sulphur shortage and the elimination of some of the bottlenecks in the existing plant at Avonmouth.

Unfortunately, however, demand for sulphuric acid has not always been so exuberant since about 1957. With increasing competition from other producers, some of this plant has fallen into disuse and Seaton Carew Works has had to be sold.

Another major item of capital expenditure in the post-war 'plan' was the decision, taken very early on in November 1945, to build an additional (fifth)

The Imperial Smelting Furnace at Avonmouth—The largest zinc
blast furnace in the world—nearing completion—autumn, 1967

lithopone pigment unit at Widnes Works at an estimated cost of £150,000. It came into production in 1950 but it was not until 1952 that it produced a product that was acceptable to the paint trade and even then only on a very limited scale. This decision has been criticized subsequently because a decline in the demand for lithopone started in 1952 and production had to be closed down completely twelve years later after every effort to stop the decline in lithopone as a commercially viable product had failed. Possibly the decision to add another production unit was taken too soon after the war, before the continuing and increasing competition from titanium dioxide and from cheaper and higher quality imported German lithopone had had time to show its strength. The new unit ran as a lithopone producer for only a short period and was later modified to become part of the plant to produce Vidox, a type of zinc oxide produced from zinc residues left after the manufacture of hydros by the dyestuffs industry.

The pre-war varieties of lithopone and 60 per cent lithopone were also re-introduced in this period, although obviously against the views of some of the Directors who considered that 60 per cent lithopone was unlikely to fulfil the hope that it would stave off the advance of titanium dioxide. However, demand for 60 per cent lithopone still continues on a modest scale in Britain, even though it is no longer made at Widnes Works. Commercial purity zinc sulphide was also brought on the sales range but never succeeded in making a break-through to large quantity sales.

The production of these high strength pigments did, however, form a chemical basis for the next development. This was that, eventually, in 1953 the Board had to face up to the fact that there might be no future for the zinc sulphide pigments produced at Widnes and introduced the production of barium chemicals from the barium side of these lithopone works, starting with blanc fixe, a high purity precipitated barium sulphate used principally as an extender in the paint and paper trades. Demand for this was small but increased sufficiently to encourage the Company to go further in this new direction. In 1955 barium carbonate and barium chloride crystals were added to the sales range and in 1964 Imperial Smelting went into partnership with Laporte, another barium chemical producer, for the rebuilding of Widnes Works for use entirely on barium chemical production in future.

The most successful chemical project on which Imperial Smelting spent large sums in these years was the development to a commercial stage of the production and sale of various types of fluorine based products. This all stemmed from the original proposal of Aluminium Sulphate Ltd. in 1939 to add aluminium fluoride to its production range. Aluminium fluoride required erection on the Avonmouth site of a small plant for production of aqueous hydrofluoric acid of 50 per cent strength, originally, as already related, by a German process licensed from Herr Schuch of Hanover; and from hydrofluoric acid Imperial Smelting's chemists have developed a wide range of products over the past quarter-century. The first hydrofluoric acid plant came into operation in 1941 and, with the aluminium fluoride plant, survived the air raid that destroyed the first aluminium sulphate plant in 1941 and continued operating throughout the war.

It appears to have been an enquiry from the Ministry of Supply which started off Imperial Smelting's major expansion into other fluorine chemicals. In June 1946 they enquired whether the Company could supply the whole or a part of their requirement for 850 tons a year of anhydrous hydrofluoric acid as existing capacity in Britain was able to supply only part of this. The capacity of the existing aqueous hydrofluoric acid plant at Avonmouth was only 160 tons (100 per cent acid) and it was occupied wholly in the production of acid for use in aluminium fluoride production. Accordingly, Imperial Smelting decided to approach The Pennsylvania Salt Company of Philadelphia, the largest producers of hydrofluoric acid and fluorides in the U.S.A., to obtain their process know-how and full engineering plans and specifications for a new plant. The agreement eventually reached with 'Pennsalt' at the end of 1946 provided for purchase and erection of a complete hydrofluoric acid plant of 500 tons a year capacity. The minute of the Board discussion of this proposal adds that 'apart from information on the production of anhydrous hydrofluoric acid the agreement with Pennsalt will provide for free exchange of information on other grades of hydrofluoric acid and fluorine products. It is believed that this exchange of information will be of considerable benefit in developing the production of fluorine products in this country for the domestic and export markets.' Erection of the plant, which had been purchased secondhand from Pennsalt and shipped over, was completed in December 1947 and it started up in February 1948. The central part of its purpose was fulfilled in the following years when the main portion of its production went to the Ministry of Supply for the Atomic Energy programme, but increasing amounts were used by the Company in pursuing Pennsalt know-how in other fluorine products and in inventing its own products. Of these one major commercial product and several minor ones established themselves on the market in subsequent years.

As early as August 1947 the Board was informed that—'Reports have been issued on the possible production of Freon by Imperial Smelting and on sources of silica suitable for the production of fluosilicic acid. Preparatory work and literary surveys connected with the production of organic fluorine compounds have been carried out and experimental work will commence shortly in discussion with the Chemical Department of Bristol University.'

These ideas were soon followed up. In April, 1948 capital expenditure was sanctioned for a project for production of elemental fluorine, sulphur hexafluoride and metallic polyfluorides. These never brought in much profit, but boron trifluoride, for which a process was worked out during the same period, established itself very quickly on the Company's sales range. It has remained there ever since and modest quantities of the gas are still sold to the Atomic Energy Authority, Harwell, and other customers. The acetic acid complexes have also been made and sold from time to time. Much energy was devoted to the construction of cells for the generation of fluorine at this stage and the help of the fluorine experts at Birmingham University, led by Professor Stacey, has been readily given from then up to the present time. In 1949 Dr. Hiscock, successor to Stanley Robson as operational chief and himself also a distinguished chemist, urged that some of the inorganic fluorides already produced on a laboratory scale should be advertised but preliminary response to publicity was disappointing as some of these were novel products in a field little known to British industry and some were already available more cheaply elsewhere. Sodium silicofluoride also established itself on the sales range at this time but the quantities sold have never been large.

The product which eventually took first place in the Imperial Smelting fluorine compound sales, Isceon, did not emerge into the news until after 1951. In that year research was stimulated by an enquiry for a supply of Freon, or alternatively anhydrous hydrofluoric acid to be used in its manufacture, in connection with a 'new insecticide spray for domestic, horticultural, or agricultural use'. Potential demand was stated to be considerable but insufficient Freon was available on the market to meet it. A few months later Cooper, McDougall & Robertson Ltd. enquired whether a chlorofluorocarbon refrigerant could be produced and it is recorded that the Research Department started looking into a new method of manufacture as royalty terms offered for existing processes were unacceptable. Early in 1952 they put up a scheme for a pilot plant based on their new continuous vapour phase process—other methods then in use were batch processes—and the continuous process later permitted the recovery of by-product hydrochloric acid. The word 'Isceon', which later became the registered trade name of the product, first appears in reports early in 1952. A pilot plant for production of Isceon refrigerants was operating by October 1952 but for some months operational and maintenance troubles prevented regular production.

It took three more years to perfect the process to the stage at which it could be operated continuously with automatic control but very rapid expansion took place in the field of fluorine chemicals after 1955. A plant to produce 500 tons a year of fluorocarbon refrigerants and propellants was built in 1956–57 and, in spite of severe price competition from the only other producer in the country at that time, had to be expanded to 750 tons capacity in 1958 and to 1,500 tons in 1959. In the same year it was found that capacity was still inadequate and it was decided to build a 3,000 tons plant and leave the 1,500 tons plant in reserve for production of newer and more experimental types of Isceon. The 3,000 tons plant was subsequently expanded to 4,500 tons and then further plant added which made it capable of producing 5,000 tons. However, both plants still had to be operated on production of the two standard grades to keep up with sales demand and a new 12,000 tons plant has recently been completed and started up. The step by step progress to this level of output has been emphasized because it reflects the cautious approach made to expansion in the face of admitted competition from other powerful producers. It proved, of course, far more expensive than would have been the riskier approach of constructing a large plant in the first place without foreseeable markets in view and, inevitably, general policy then tended to react to the opposite extreme. As a result of this reaction, the latest zinc smelter expansion scheme has been framed on a gigantic scale which should provide for most foreseeable eventualities in increase in demand for some years to come.

The continuing researches of Imperial Smelting's able team of fluorine experts and consultants added a further range of fluorocarbons to the sales range recently in the form of two carbon Isceons. In addition, at the end of 1958, progress was announced by the Research Department in introducing fluorine for the first time into the aromatic range of organic compounds, thereby producing for sale a new set of intermediates known as highly fluorinated aromatic compounds, which require the further researches of manufacturers to produce useful end-products. These products are now on the verge of a significant break-through although the development is still in its early stages.

This has been a lengthy catalogue of old processes and new products which even now has had to omit some of the lesser sidelines of the industry in these years such as metallic arsenic, vanadium catalyst, 'C-Sentry' anodes for ship protection and the production of acid from ferrous sulphate residues. An account of these would reinforce the impression that the previous pages may give of the unco-ordinated and opportunist pursuit of established and likely profit making lines, which is probably as true of Imperial Smelting in the post-war years as it is of most companies at any time. Conscious long-term planning, in the form of the separate 'skill' that it has now become, was in its infancy throughout much of this period and several of the new projects and the particular emphasis given to policy in respect of some of the older products, such as lithopone, leave an impression of personal inspiration at work behind the scenes declining at times into opportunism among the personalities who have sat in conference, and sometimes in discord, at the top.

The great project outlined in the next chapter was also a cause of disunity in the darkest and most expensive years of its development but has abundantly justified in the long run the optimism of those who had faith in the scientists' dream from which it grew.

CHAPTER FOURTEEN

1937–1967
The Advent of
the Imperial
Smelting Process

by S. W. K. MORGAN

BRIEF HISTORY OF PREVIOUS PROCESSES—THE AWKWARD
METALLURGY OF ZINC—THE REDUCTION OF ZINC OXIDE
PREVENTION OF THE REVERSE REACTION—THE HISTORY
OF THE CONDENSATION PROBLEM—THE LIMITATIONS OF
RETORT METHODS—THE 'THIRTIES AND THE FIRST LOOK
AT THE BLAST FURNACE IDEA—THE KEY PROBLEM OF THE BACK
REACTION AND ATTEMPTS TO SOLVE IT—THE SOLUTION AND
HOW IT WAS TESTED—THE PROBLEMS OF TRANSLATING
THE NEW PROCESS INTO A PILOT FURNACE—THE EXPERIMENTAL
BLAST FURNACE STARTS UP—HITHERTO UNREALIZED
POTENTIALITY AS A SIMULTANEOUS SMELTER OF BOTH
LEAD AND ZINC—WOODS AND LUMSDEN WRITE UP THE SCIENCE
OF THE NEW PROCESS—DECISION TO DESIGN A COMMERCIAL UNIT
THE BOARD STARTS TO TAKE THE PROJECT SERIOUSLY
RELATIVE ECONOMICS OF THE OLD PROCESSES AND THE
NEW STUDIED—DECISION TO BUILD TWO COMMERCIAL UNITS
INSTEAD OF ONE—START-UP OF THE NEW FURNACES
ONLY LIMITED SUCCESS—BOARD ROOM PRESSURE
IMPROVEMENT OF CONTINUITY AND SINTER
THE PROCESS AT LAST BECOMES AN ECONOMIC PROPOSITION
DECISION TO DESIGN LARGE-SCALE FURNACE FOR SWANSEA
RAPID SPREAD OF IDEA TO OTHER COUNTRIES
THE NEW ENLARGED AVONMOUTH FURNACE—VACUUM DEZINCING,
REFLUXERS AND THE UPGRADING OF THE METAL
SIGNIFICANCE OF THE WHOLE DEVELOPMENT FOR THE
BRITISH AND WORLD ECONOMIES

TO UNDERSTAND the importance of this development in the world history of zinc smelting, it is appropriate to begin by summarizing much of what has been related in this book about the technical aspects of producing zinc.

The National Smelting Company was formed to produce zinc by the horizontal retort process and this process, with its disadvantages of high labour and fuel costs, was the main source of zinc production in the Imperial Smelting Group until the last furnace was shut down at Swansea in 1960. The vertical retort process, operation of which was first commenced in 1934 and which still continues, was certainly an improvement, since it was more mechanized and the cost of fuel and labour was reduced. It was still a retort process, however, and had the unescapable disadvantages inherent in this method. Even in the early days of the Research Department dissatisfaction with the horizontal process was acute and this feeling grew steadily as the process continued to resist all efforts to improve its performance. Out of this frustration the zinc blast furnace was born.

To explain the origins of the blast furnace process, it is necessary to recall a little elementary metallurgy at this point.

From the metallurgical point of view, zinc has awkward properties and is not an easy metal to produce. It occurs in nature largely as zinc sulphide, and is generally associated with lead, silver, iron and sometimes copper. It is concentrated at the mine by the flotation process and for orthodox smelting processes great efforts are made during flotation to produce a concentrate, consisting largely of zinc sulphide.

Zinc sulphide, however, is an unresponsive material. It is relatively inert chemically, and the first step in all extraction processes is to convert it into zinc oxide, which is more reactive and will respond to treatment. As opposed to the sulphide, for instance, zinc oxide can be dissolved in sulphuric acid. The solution of zinc sulphate so formed can be electrolyzed after extensive purification and metallic zinc deposited. This forms the basis of the electrolytic process but even with this process zinc displays idiosyncrasies. Theoretically it should not be possible to deposit zinc from sulphate solution by electrolysis at all as hydrogen should be produced instead. That this does not happen is due to an imperfectly understood phenomenon called hydrogen overvoltage. It can, but perhaps should not be, argued that the electrolytic process, which at the present time produces over half the world's zinc supply, depends basically on what may be termed a quirk of nature.

However, Imperial Smelting has never been concerned with the electrolytic process since the cost of electrical energy in Britain has always been prohibitive. Therefore it has had to use smelting methods based on another property of zinc oxide—the fact that it can be reduced by carbon.

Here again there are considerable difficulties. Zinc oxide is a stable oxide, and much more energy must be used on its reduction than is the case with other common metals such as lead and copper. Before reduction can proceed by either carbon or carbon monoxide, high temperatures—of the order of 1,000°C—must be used. Another complication arises at this stage because zinc metal has the relatively low boiling point of 907°C. Thus at ordinary pressures the result of the reduction reaction is to produce zinc in vapour form, and this must be condensed before liquid or solid zinc can be obtained. At this point the greatest difficulty of all is encountered, because, as the gases leaving the reduction zone containing zinc vapour and carbon dioxide are cooled to effect condensation, the reverse of the reduction reaction occurs rapidly, and the end result tends to be a mixture consisting mainly of zinc oxide and carbon monoxide, i.e. largely the initial starting point.

The problem of condensation has dominated the whole history of zinc. The early metallurgists could produce copper, tin, lead and iron, and developed hearth and blast furnaces to do so, but zinc could not be smelted by these means. As described in Chapter 1, the first solution of the condensation problem seems to have appeared in India or China in the fourteenth and fifteenth centuries, when a number of small retorts, each with a separate condenser, were piled in a heated furnace. Each retort was separately charged with a mixture of calcined zinc carbonate and charcoal and heated to a high temperature. The zinc oxide was reduced by the charcoal and left the retort as vapour, to be condensed partly to liquid metal in the external condenser. Few details of the construction of the furnaces are known. They must have been small and inefficient but they were almost certainly the first known producers of metallic zinc.

A description has also been given in an earlier chapter of Champion's achievement in producing zinc at Warmley, near Bristol, for the first time in Europe, but this was not until the early eighteenth century. His process was an improvement over what little is known of the earlier Chinese or Indian practice, in that he used much larger crucibles which would be somewhat easier to charge and discharge. It was still a retort process, however, with a very high labour cost and poor thermal efficiency.

The Champion or English process remained the only process for producing zinc until the end of the eighteenth century. Then a big step forward was made in Belgium, where furnaces were built using horizontal clay retorts fixed in rows, each with its own fish shaped condenser to condense the zinc. Every condenser was removed at the end of each distillation cycle to enable the retorts to be discharged and then recharged. This basically was the Belgian or

M

horizontal process, which remained fundamentally unchanged for the next 150 years. Although the last horizontal furnace was shut down at Avonmouth in 1958 and at Swansea in 1960, it is still in use in various parts of the world, particularly in the United States.

A big improvement in retort practice was made in the late 'twenties by the New Jersey Zinc Company in the U.S.A. with the development of the vertical retort process. Reference has already been made to this process which solved for the first time the problem of continuous distillation. Whilst with the horizontal process the maximum size of retort which could be used produced only 80 lbs. of zinc per day, the original New Jersey vertical retorts could produce $2\frac{1}{2}$ tons which, with gradual improvement over the years, has now been raised to over 7 tons per day. Perhaps the greatest achievement of the New Jersey Company, however, was that they developed an efficient condenser to deal with the relatively large volumes of zinc evolved from the retorts.

In the 'thirties a somewhat parallel development was produced by the St. Joseph Lead Company which again enabled large distillation units to be built. A mixture of coke and sintered zinc blende was fed into a large retort furnace. This, however, was not heated externally as with the other retort processes, but internally, through electrical energy making use of the resistance of the coke in the mixture to raise the temperature of the charge to the necessary levels. The fresh factor introduced was a new type of condenser. By reducing the pressure the zinc-containing gases leaving the top of the furnace were sucked through a bath of molten zinc held in a cooled vessel and thus efficient condensation was obtained.

This brief description of the Electrothermic or St. Joseph method of producing zinc hardly does justice to the process which is operated very efficiently at Josephtown in the U.S.A. where units producing over ninety tons of zinc per day have been built; but, for reasons which are not clear, it has had little application elsewhere.

From the analysis of the development of the metallurgy of zinc given in this book, it can be seen that, before the blast furnace work at Avonmouth, all previous thermal processes for producing the metal have been forced to use retort methods owing to the difficulties posed by the condensation problems. A retort process has certain disadvantages which are difficult to avoid. Because the large amount of heat necessary to complete the reaction must be forced through a retort wall, the process tends to be thermally inefficient and restricted in the size of unit which can be built. Although the electrothermic process largely overcomes these particular drawbacks, it does not escape what is the most serious disadvantage of a retort method—that the charge cannot be permitted to fuse during reduction but must remain in a 'dry' condition so that it can be readily removed from the retort. This means that the temperatures employed must be limited and only relatively high grade materials can be used as raw material for the process.

It was obvious that these difficulties, which have had such a cramping influence on the whole development of zinc metallurgy, would be overcome if a blast furnace process could be used. The vital question was whether or not the condensation problem could be solved.

A detailed study of the possibilities of blast furnace production of zinc began in the Avonmouth Research Department in 1937. Search of the literature did little to stimulate optimism. A number of attempts had been made in the past but all had ended in expensive failure. None had succeeded in condensing zinc as metal but only as a dust or powder. Even the pundits were pessimistic. In 1933 C. G. Maier of the U.S. Bureau of Mines had published a study of the thermo-dynamics of zinc smelting which was a landmark in the scientific development of the industry. Although, as the first analysis of its kind of the basic conditions of zinc oxide reduction, it was most valuable, Maier came to the conclusion that blast furnace production of zinc was quite impractical and thus scholarship lengthened the odds against the project.

Nevertheless, Imperial Smelting scientists were not completely daunted by the weighty arguments against the possibility of ultimate success. The key problem was obviously the dreaded back reaction between carbon dioxide and zinc vapour. In retort operation this could be avoided largely by keeping a low proportion of carbon dioxide in the vapours leaving the retort but with a blast furnace very much higher proportions of carbon dioxide were unavoidable and the problem was intensified. In spite of all this a belief persisted, albeit somewhat vaguely, that it should be possible to cool the gases leaving the furnace at such a rate that the back reaction would have insufficient time to take place so that at least some liquid zinc could be produced.

Some encouragement was gained during a visit to Baelen in 1937 to examine the operation of a shaft furnace, built there to produce zinc oxide from electrolytic plant residues. Although air was added to the gases after they had left the furnace, to burn them deliberately, it had been found necessary also to add some steam inside the furnace to prevent the formation of accretions, which were found to contain some lump metallic zinc. Was the back reaction therefore as powerful as it appeared?

In 1938 it was decided to start some experimental work at Avonmouth, primarily to examine the effect of rapid cooling during condensation. Implementation of this programme was delayed as the efforts of the metallurgical team had to be concentrated on the more urgent problems of increasing the output from the vertical and horizontal retort furnaces during the first years of the war.

By 1943, however, it was decided that a limited amount of work could be permitted and a small furnace was built into which air was blown both at the top and the bottom. The gases were withdrawn at a point halfway down the shaft and then entered a vertical column packed with coke. These unconventional precautions were taken in the belief that by adopting them re-oxidation of zinc by carbon dioxide would not occur until the very last moment. After emerging from the top of the coke column the vapours entered immediately a chamber filled with tubes cooled internally with water so as to give the most rapid cooling then possible.

Some promising indications were obtained at once. It was obvious that zinc could be eliminated from the charge and that quantities of zinc dust high in metallic content could be made; but it was doubtful whether significant progress had been made beyond the point reached by predecessors in this field and the goal of a continuous process producing liquid zinc still seemed far away.

Then a revolutionary suggestion was made by L. J. Derham who had been associated with the project since its start. His proposal, which transformed the whole problem, provided the vital key which was to lead to eventual success. However, many weary years of concentrated effort by a large part of the Avonmouth staff were still to elapse before it was proved that a fully commercial process had been evolved.

Derham's suggestion was to quench the vapours leaving the coke column in a dense shower of droplets of molten lead thrown up continuously into the gas stream as it passed through the condenser. By this means the vapours would be shock chilled and the temperature reduced almost instantaneously, so that the back reaction would have no time to occur, until temperatures were reached at which it was innocuous.

Lead was an almost perfect choice of medium. It could be readily handled in steel with the result that steel rotors could be used to produce the spray inside the condenser and steel pumps to circulate it through an external cooling system. What was even more important, while it chilled the vapours inside the condenser its own temperature rose and it dissolved the zinc as it formed. The hot lead could be withdrawn from the condenser and passed through an external cooling system. The zinc as it cooled became less soluble and separated as a liquid phase on top of the molten lead. It could be removed by liquation and the cooled lead, now stripped of some of its zinc, could be returned to the condenser to repeat its shock chilling effect. A continuous process was thus possible.

It was decided to test this concept at once. The original water tube condenser producing zinc dust was discarded and a new condenser was built consisting of a chamber with a pool of molten lead four inches deep in the bottom. The spray was produced by a vertical paddle wheel, which dipped in the molten lead and, when rotated, threw up large quantities of spray partially filling the whole volume of the condenser. By this time the project was overspent, a condition which was to persist for most of the next twenty years, and the Research team could not afford a motor to turn the paddle wheel but had to rotate it by hand. For the same reason the charge also had to be hand fed to the furnace in buckets.

On 21 May 1943 the furnace was operated with the new condenser for the first time and encouraging results were obtained almost at once. As long as the paddle could be rotated at an adequate rate, it was obvious from the appearance of the exit flame that zinc was being removed from the gas. Although the trial lasted only six hours, over 50 lb. of zinc were recovered from the condenser. A number of further trials were run during 1943 and the promising results were repeated, but then, due to pressure of other commitments, the work was once more held in abeyance.

Between 1943 and 1945 no practical work was done on the project but a considerable amount of thought was given to what should be the next stage. It was realized that little more could be gained by working on the previous scale, with makeshift equipment. This had served its purpose well in demonstrating the vital principle that, if the vapours leaving the furnace were shock cooled by dowsing them with an intense spray of lead droplets, the back reaction could be held in check and molten zinc produced. The next step obviously was to prove that this could form the basis of an operating process, and this could only be done with properly built equipment. In particular a certain amount of mechanical handling was essential, together with a fair degree of instrumentation. Two main difficulties had to be overcome at this stage—to design the pilot furnace and equipment which would be required, and to prepare the Board for the considerable capital expenditure involved. The latter proved to be the harder task.

Even at this early stage several requirements for successful operation were realized to be essential. It was obvious that if preheated air and coke could be used this would reduce the coke consumption per unit of zinc. A furnace was designed, therefore, which was rectangular in section, of internal dimensions 2 ft 6 in. by 3 ft 6 in. by 10 ft high and fitted with two tuyères at the bottom, through which air could be blown, after heating to a temperature of 450°C in a tubular air preheater. A mechanical hoist was to be used to elevate the charge, which consisted of lump sintered blende together with coke preheated in a separate internally fired shaft and the charge was fed into the furnace through a manually operated bell. At first the gases were withdrawn half way up the furnace, and passed up through a column of hot coke attached to the main shaft, and from the top of this column into a condenser which consisted of a rectangular chamber, fitted with a vertical paddle wheel partially submerged in a pool of molten lead. When the wheel was rotated a shower of lead droplets was formed through which the gases had to pass. If the indications that had been obtained from the earlier work were correct, this spray would chill the gases and condense and dissolve the contained zinc. A vessel fitted with air cooling apparatus was built alongside the condenser through which lead drawn from the pool could be circulated and cooled. As it cooled, molten zinc came out of solution and, being lighter, floated on top of the lead from which it was removed by skimming. The cooled lead was then returned to the condenser.

This was to be the design of the first experimental blast furnace. It was intended to produce two tons of zinc per twenty-four hours, but it was several years before it achieved this target. Its initial cost was estimated to be £13,750, and the next problem was to use the flimsy evidence supporting research optimism, to obtain permission to spend this sum on a project which many metallurgists at this time would have considered to be impossible. It is much to the credit of the Board that they saw the potentiality of the process and agreed to the expenditure, so that construction started immediately. The furnace was completed by mid-1947 and operations commenced.

The project was complex and expensive from the start. Each shift had to be manned by ten operators and controlled by technical staff. Almost every part of the equipment gave trouble. Stoppages were frequent and progress was painfully slow; but even in the first days of operation some metallic zinc was made and it was obvious that the basic principles were sound. The problem was whether the many operating difficulties could be overcome, and a process capable of commercial operation developed. This remained open to doubt for the next ten years.

During the early days of operation mistakes were made as the staff and operators learned by trial and error the new techniques demanded by the process. Nevertheless two important simplifications were effected at this stage which gave a real gain in practicability. The coke column, which was originally installed to condition the gases before condensation, was found to be unnecessary, and the paddle wheel installed to spray the lead in the condenser was replaced by more practical vertical impellers inserted through the roof.

Considerable gains were also made on the metallurgical side. One of the early fears was the doubt whether zinc oxide could be reduced to the necessary degree in the furnace, whilst leaving the iron oxide in the charge unreduced to form slag. Had it been necessary to reduce large amounts of iron oxide in order to obtain a high degree of zinc elimination, not only would the coke consumption have been excessive but, in addition, very high temperatures would have been necessary at the furnace bottom in order to remove the iron in a molten condition. Experience showed, however, that, with a suitable choice of charge/coke ratio, the zinc oxide could be reduced almost completely leaving the iron oxide unchanged, to pass into the slag.

At the same time a discovery of great significance was made. The process was originally conceived as a zinc process alone. With the use of lead in the condenser as a shock chilling medium a certain amount of lead-containing dross was formed and, in order to recover this and the zinc associated with it, it was returned and added to the charge on the sinter machine. Gradually the lead content of the sinter rose. As it did so lead began to appear with the slag at the taphole, and could be separated and recovered. It therefore became apparent that the furnace could be considered an efficient producer of metallic lead as well as of zinc.

This discovery should not have been the surprise that it was, as further consideration soon showed that the reduction of lead oxide by carbon monoxide, which must occur at the top of the furnace, was an exothermic reaction and thus did not absorb heat from the furnace but actually contributed to it. The reduction of quite large quantities of lead oxide, therefore, should not lower the capacity of the furnace to produce zinc, and there was justification for the claim that lead 'got a free ride'. Although an over-simplification, this statement is largely true and represents one of the greatest advantages of the process.

The 'free-ride' concept also meant, if it was true, that lead concentrates could be smelted cheaply in the furnace, and thus the economic recovery of both lead and zinc simultaneously was possible. An ancillary benefit arose from the fact that lead is a good solvent for silver and other precious metals in the charge, and also for copper. These metallic values were thus extracted and carried out with the lead from the furnace bottom as bullion, from which they could be recovered by standard lead refinery practice.

This development was important in its own right but had implications beyond the furnace itself. At many mines zinc and lead sulphides occur together with small amounts of silver and sometimes of copper. In many cases separation by differential flotation is not difficult and separate zinc and lead concentrates can be made at good recovery. At a number of mines, however, efficient separation is difficult, if not impossible, and individual concentrates cannot be made without incurring heavy losses of values. Once the capacity of the new blast furnace to treat zinc and lead together was established, it was no longer necessary to produce separate concentrates at the mine. In a number of cases a single bulk concentrate could be made containing all the values, since this formed an ideal feed for the new furnace. Advantage is now taken of this possibility at a number of mines.

This period of development of the process consisted largely of a struggle for survival against the practical problems which threatened to overwhelm it. A considerable amount of work was done, however, on a number of aspects of the process. A suitable design of rotor to develop the lead spray in the condenser was evolved. At the same time the thermodynamics and general physical chemistry of the principles involved were subjected to an exhaustive study. This culminated in the production of a report by Dr. S. E. Woods and John Lumsden which stated clearly the basic chemistry of the process, and placed its operation for the first time on a logical basis. These two had joined the staff of the Research Department in 1935. Both were first class physical chemists and together provided a firm theoretical basis for most of the subsequent work of the department. Lumsden specialized particularly in thermodynamics and became an expert of world-wide renown in this field and the author of two standard books on the thermodynamics of alloys and of molten salts.

In January 1948, a big step was taken. A considerable amount of work had by now been carried out on the experimental blast furnace, which was beginning to operate more regularly and produce results with greater consistency, so it was decided that sufficient progress had been made to justify designing a larger furnace as the first commercial unit. With the clarity of vision which hindsight always brings, it can be argued that this decision was premature, but confidence by now had spread from the Research Department to the Boardroom.

The Board were beginning to give this project the highest priority at this time when 'The Chairman referred to the potential possibilities of the blast furnace project and asked the Board to treat any information which may be submitted as strictly secret and confidential'. Five months later, in June 1948, they approved in principle the recommendation of Digby Neave, the Managing Director, for 'proceeding with the erecting of the first of six large size improved vertical furnace units on a site at Avonmouth Works where they could, if found desirable, at a later date replace the horizontal retorts'. A 'Capital Budget' appearing in the Board papers later in 1948 shows the intention to erect six of these furnaces at Swansea also at a total estimated cost of £875,000 for each site.

Possibly the management were thinking in those days of the new furnaces being not much larger than the existing vertical retorts but, if so, this conception was soon changed after observation of the potential output of the first two commercial size furnaces a few years later. Formal approval was given in January 1949 for the building of the first furnace at an estimated cost of £309,000 including foundations and buildings.

It was laid down that this first large unit should have a capacity of twenty tons of zinc per day, which was some ten times greater than the experimental blast furnace had achieved when operating at the peak of its form. In the attempt to reach this target, a rectangular furnace of a hearth area of sixty square feet was visualized. As with the experimental furnace the gases were to be taken off half way up the charge by means of a tunnel flue running across the furnace and leading into two condensers, one at each side. Each condenser would be divided into three compartments fitted with vertical rotors. Hot lead containing dissolved zinc was to be withdrawn from the condensers and cooled by air in an external vessel, one of which would be fitted to each condenser. As the lead cooled, the zinc should come out of solution and float on top of the lead from which it could be run off into a holding bath and cast as output.

The plant was to be highly instrumented with the maximum of automatic control, a special feature being an intricate system which governed the pressure in the condensers and the volume of furnace gas which each was to handle. This proved highly successful and was an integral part of all two-condenser furnaces subsequently built.

From this time on until the late 'fifties, when the horizontal distillation plants were gradually closed down, consideration of policy on the new process became closely linked with consideration of the changing economics of the old process, as mentioned in the previous chapter. In April 1949 review was made 'of the current advantages of a more modern process'. There were as yet no reliable data on which to make an estimate of the running costs of a zinc blast furnace but a comparison between the Swansea horizontal distillation plant and the Avonmouth vertical retorts showed a £10 a ton advantage in favour of the vertical retorts with the American zinc market price currently at 12 cents a lb., and falling at that time. The devaluation of the Pound against the Dollar on 19 September 1949 sent the Government controlled zinc price up from £63. 10s. 0d. to £87. 10s. 0d. a ton and removed disquiet about horizontal distillation economics for the time being. The horizontal retorts, of course, made good profits during the Korean War price boom but there are frequent references in the records to low recoveries and the ineffectiveness of all measures tried for improving the position. There are continuing references also to labour shortages and labour troubles caused by the depressing effect of the plant results on the men's wages. More specifically, when the zinc price fell during the second half of the Korean War, G. R. Daniel, the Deputy General Manager, concluded a report on the economics of the situation with the statement that 'the horizontal process, as at present operated, could not hope to survive a greatly lowered zinc price'. This thought, backed up by comparative costs statements, was constantly before the Board in the next few years although there were bursts of optimism in the opposite direction whenever the zinc price rose. Not only were the horizontal distillation furnaces becoming more and more uneconomic but it became increasingly evident also that it was impossible to improve their performance to the degree necessary for their survival. Accordingly the conclusion was reached that the future of zinc smelting in the Company depended on the success of the blast furnace project and that it was logical therefore to build not one, but two furnaces, with a combined capacity of forty-five tons per day, which would thus enable the horizontal furnaces to be shut down. This was a courageous decision and much credit must be given to the Board for taking it, although the wisdom of this step was frequently to be challenged by the troubles which lay immediately ahead.

In December 1948, therefore, work began on designing a second furnace before the construction of No. 1 had been completed. The design was basically the same as that of the earlier unit but it was slightly larger in size—70 square feet hearth area—and was expected to produce twenty-five tons of zinc per day. It was estimated to cost only £203,000 as it would use some of the No. 1 furnace facilities, and permission to proceed with its construction was received in May 1949.

On 28 September 1950, No. 1 furnace was blown in. The first campaign lasted only six days and zinc production averaged 4·2 tons a day. It was a good start but many plant difficulties were experienced, particularly with the charging system, the gas washing equipment, the coke preheaters and the refractory lining of the condenser which was attacked by carbon monoxide. All these required modification. One of the most serious deficiencies lay in the coolers provided to cool the lead withdrawn from the condenser and they had to be discarded and replaced by a type of water-cooled launder which had been used successfully at Port Pirie in lead refining.

As continuity improved, other weaknesses revealed themselves, particularly with the materials of construction of the rotors and their shafts, the hot end of the furnace charging gear, and the furnace roof, and daily cleaning of the tunnel flue became essential.

An intensive campaign to overcome these troubles was mounted and all the facilities of the Works Engineering Department were put at the disposal of the project. Indeed, one of the most encouraging features at this time was the co-operation freely given by everyone on the works' site. As a result of this drive, operation improved and in the first 370 days of operation the furnace was on blast for 243 days and produced 2,069 tons of zinc. It was still painfully obvious, however, that a satisfactory operating process had not yet been developed.

By this time increasing pressure was being brought to bear from the Boardroom for the project to be advanced to a commercial level of efficiency at the earliest possible date. The target date set by Govett, the Chairman, had been 31 December 1949. He was worried not only by the economics of continuing zinc production from the horizontal retorts but also by the Australian situation. For Commonwealth political reasons great emphasis was being placed at that time on renewing the drive for the industrialization of Australia. Neave had gone on a fully publicized mission to Australia in February 1948 in connection with projects for 'new plants for the production of sulphuric acid from concentrates, utilization of this in fertilizer and commercial use of zinc metal'. Govett stressed to the Board in March 1949 the vital importance of co-operating in the programme for the treatment of a greater quantity of Australian concentrates in Australia than were being used by Electrolytic Zinc Industries at Risdon alone as a factor in securing the continued co-operation of the Australian Commonwealth and State Governments in overseas shipments of Australian concentrates. Plainly Imperial Smelting must be in a position to offer the Australians the new process as soon as possible. Every effort, therefore, was made to expedite the building of these furnaces, which was in the hands of contractors, but delays still occurred on both furnaces.

The second furnace at Avonmouth did not start up until 2 April 1952. Part of the delay was due to alterations made in the course of erection in order to incorporate improvements suggested after over a year's operation of No. 1 furnace. However, as soon as two furnaces were in operation it was decided that No. 1 could now be used as an experimental unit to attack problems which were still only too evident.

Arising out of this experimental work it was decided to abandon yet another belief cherished from the early days of the work. There had been an established conviction that, in order to attain the best conditions for satisfactory condensation, it was essential to withdraw the gases half way up the furnace—hence the use of a tunnel flue but, as the scale of operation grew, this became the major cause of discontinuity. As a result of continuing experimental blast furnace studies and tests on No. 1 furnace, a decision was taken to convert No. 2 to bottom blowing with a gas offtake at the top of the furnace. This was carried out at the end of 1952 and was immediately so successful that it was obvious that a major step forward had been taken.

Many problems still remained. The furnaces had already shown themselves capable of producing zinc at reasonable efficiency at rates much higher than those originally foreseen but the main outstanding difficulties still lay in the maintenance of operation. Until the furnaces could be made to operate smoothly over long campaigns the practicability of the process was still unproved. The next phase of the work therefore was directed towards the vital question of improving continuity.

Almost immediately theoretical reasoning proved helpful. There was now no doubt that the back reaction—the re-oxidation of the zinc vapour on cooling —could be held in check by the lead splash condenser, and metallic zinc produced but, before the gases could enter the condenser, they had to emerge from the furnace top. Here they had no protection and heavy deposits of zinc oxide were formed on the furnace roof and condenser inlets. Every four days the furnace had to be shut down for removal of these accretions, and even running was impossible. A number of possible palliatives were tried, such as the use of a heated flue connecting the furnace to the condenser, but little satisfaction was obtained until an elegant solution was put forward by Dr. Woods.

The gases emerging from the top of the furnace charge contain zinc vapour, carbon dioxide, carbon monoxide and nitrogen. If the temperature is maintained, they continue in equilibrium, but if the temperature drops, then immediately carbon dioxide begins to re-oxidize the zinc vapour, which is, of course, the back reaction about which so much has been written. Woods pointed out that, if air is added to the mixture of gases, some of the carbon monoxide burns, forming more carbon dioxide. This increase in carbon dioxide concentration will have an adverse effect on equilibrium, but the increase in temperature contributed by the burning of carbon monoxide more than compensates for this. The net result is, therefore, to produce a mixture which can stand a certain degree of cooling before zinc oxide begins to form and thus some protection is given against re-oxidation before the condenser is reached. Thus one reaches the superficially contradictory solution that to prevent oxidation one should add air.

The principle was tried out on No. 1 furnace and gave encouraging results immediately. Work had to be done to determine the best method of admitting the air and the optimum amount to admit, but as soon as these points had been determined it was obvious that a real contribution to the problem of continuity had been made, and additions of air to the furnace top are now an essential part of the operation.

Although the principle of top air addition made a valuable contribution to the problem of maintaining continuity, further improvements were essential, as the operation of the furnaces was still far too variable. At peak performance the results were very promising but the periods of sub-standard operation were still much too long. The cost of the frequent modifications were heavy, and, in addition, due to the low average performance, running costs showed a heavy loss. It was estimated that, by 1957, the company had already spent over £3 million on the project but the process was still not completely proved. Whilst confidence still ran high amongst those closest to the operation, it was only natural that alarm should begin to spread amongst those responsible for the finances of the Company. It was a difficult period.

At this stage, nevertheless, although their performance continued to be uneven, the furnaces were at least beginning to work. The next task was to raise their average performance to economic levels, which is an essential, if unspectacular, phase in the development of all processes, and is one for which sufficient allowance is rarely made. Consequently, it is in this period, which is of necessity the most expensive, that most projects founder.

As part of the drive to obtain even economic operation, attention was directed to improving the charge since it was realized, soon after operations began, that this was far from ideal. From the outset the new process had demanded a sinter which was much lumpier and harder than that required by the horizontal or vertical retort processes, to which the roasting operations of the Company had previously been geared. No. 1 downdraught machine had been allocated to produce sinter for the blast furnace, and immediately considerable effort and money had to be expended on improving the crushing, screening and cooling circuits. This proved insufficient. As the need grew to treat more lead, the downdraught machine became more difficult to operate. During sintering lead tended to migrate down through the bed and collect on the grate bars, which it attacked with a ferocious intensity. The problem became so serious, as the shutdown periods on the machine increased in length, that the machine began to be unable to supply the furnaces with their full demand for sinter.

Fortunately, a solution to this problem was already available. The problem of grate bar attack had been serious in lead smelting for a number of years, but it had been brilliantly solved at Port Pirie and elsewhere, by reversing the flow of air and blowing from below. The gases containing sulphur dioxide were then withdrawn from above the bed by means of a close fitting hood to which suction was applied. In these conditions metallic lead was contained in the bed and did not reach the bars, which were also protected by the cooling effect of the incoming air.

It was decided to adopt this solution but it meant a very radical change in operations and, although Imperial Smelting were fortunate in being able to draw freely on the Port Pirie experience, it was soon found that conditions differed from theirs, and a new crop of problems appeared.

The first arose from the fact that the density of the zinc-lead concentrate mixture which had to be treated was appreciably less than that of lead sulphide alone. The tendency of the bed to lift and form blow holes under the influence of the blast was correspondingly greater. This necessitated paying extreme attention to the preparation of the charge—good sizing, proportioning, and mixing being more critical than with lead sintering. Every effort had to be made to reduce densification of the charge, and to avoid compaction as it was fed on the machine. Also ignition was more difficult than with lead sintering, and this imposed special problems at the feed end of the machine.

The vital decision to convert No. 1 machine to updraughting was taken in May 1955. It was a big step and there was a painful period of two years during which the conclusions outlined in the previous paragraph were realized and implemented. It was, in fact, during this period that the fate of the whole blast furnace project hung most precariously in the balance.

However, success was achieved eventually. The quality of sinter production vastly improved and it became possible to produce for the first time a hard, suitably sized feed for the furnaces. These responded to the better diet which could now be provided and, with the modification described above, which had been made to the furnace at the same time, the continuity and average standard of operation improved greatly, and for the first time there was real evidence that the process was an economic one.

One further major modification in furnace design was made shortly afterwards. In 1956, as the output from the furnaces began to increase, the sintering capacity became inadequate to feed both furnaces and No. 1 furnace was shut down. While it was out of commission, it was rebuilt with a wider shaft of 107 square feet in area. No alteration was made in hearth dimensions and the increased width of shaft was obtained by using the chair or stepped type of jackets which had been developed at Port Pirie. This proved to be a successful move and the output from No. 1 furnace rose to over sixty tons a day. No. 2 was promptly rebuilt in the same way and the condensers were enlarged. No. 1 was shut down permanently in 1959 but No. 2 continued in operation until November 1967 when it was shut down to make way for the large No. 4

furnace recently completed at Avonmouth. No. 2 furnace was, until late in 1967, regularly producing ninety tons of zinc and fifty tons of lead a day, and some measure of the improvement which has been made can be gauged when it is remembered that it was originally designed to produce twenty-five tons of zinc a day.

The beginning of 1958 clearly marked the end of an era in the development of the blast furnace process. It no longer needed the eye of faith to see its potentialities. The performance of the improved No. 2 furnace was now such that even the cold analysis of the accountant showed that the process had at last arrived and that a new factor had arisen in the zinc industry.

With the evidence now available, a decision to take the next step forward was made and, early in 1958, it was decided to design an even larger furnace to gain the maximum economies of scale. An increase in capacity of approximately 100 per cent seemed justifiable, and design work began on a furnace with a shaft area of 185 square feet.

This furnace had all the features which had been developed and established on Nos. 1 and 2 furnaces and a number of further refinements were introduced. Air preheat temperature was raised to 750°C, more extensive automatic control was applied, and the accuracy of charge proportioning was increased.

The first of these 'standard' 185 square feet furnaces was built at the Swansea Works of Imperial Smelting. The horizontal furnaces there were shut down over 1959 and 1960 and the new furnace was first commissioned in March 1960. A second 'standard' size furnace was almost immediately commissioned for Sulphide Corporation at Cockle Creek in Australia. A big and highly successful change was made in the design of this furnace, in that it was decided to use one condenser instead of the two which had been provided for the Swansea furnace. Since then others have followed and, at the present time, standard furnaces of the Swansea type have been adopted by other overseas companies and are now in operation in the following further places:

France	Noyelles Godault
Zambia	Zambia Broken Hill
Rumania	Copsa Mica
Germany	Duisburg
Japan	Befu
Canada	New Brunswick

Furnaces are also being built in Poland, Japan (a second one), Sardinia, and in China.

It is possible that the final stage is now being reached at Avonmouth. A furnace has been built there with a shaft area of 292 square feet. This furnace should eventually be capable of producing 120,000 tons of zinc and 80,000 tons of lead a year, which must represent about the ultimate in single unit furnace capacity at the present time. The furnace is due to be commissioned at the end of 1967 and represents the climax of what has been a long and arduous period of development. To enable this new plant to operate with the maximum of efficiency, Avonmouth's existing sintering, cadmium, and acid plants, which are mainly of 1917–34 vintage, have been replaced by modern equipment, taking full advantage of the opportunities for computer control. The increased production of sulphuric acid will be sold mainly for fertilizer use, through the production of phosphoric acid.

The scheme as planned takes account of the fact that, as mentioned earlier in this book, the trend of usage has moved in the past twenty years towards the purer grades of zinc. The Imperial Smelting Research Department have in recent times worked out a technique for upgrading zinc of 98·5 per cent purity to zinc of over 99·9 per cent purity by a method based on vacuum distillation. A plant for production of commercial quantities of metal upgraded by this method has already been added to the Swansea blast furnace. At Avonmouth, however, it is intended to revert once more to the refluxing technique in order to supply part of the large quantities of 99·99 per cent purity metal required for Imperial Smelting's production of Mazak zinc alloy, which is at present wholly dependent on imported zinc.

The overall effect of the whole scheme will be a significant contribution to the national economy resulting from a reduction in the amount of brimstone for sulphuric acid production and of zinc metal that Britain has to import at present to meet the balance of home demand. It is calculated that, when the scheme comes into full operation shortly, the zinc industry's contribution to the national balance of payments will rise from £5¼–£6½ million a year to £8½–£10¼ million a year depending on the level of zinc and brimstone prices.

Some idea of the commercial growth of the process can be obtained from the fact that, by the middle of 1967, over a million tons of zinc had been made by the process and, at the present rate of progress, it is estimated that by 1975 over 25 per cent of the world's zinc and 18 per cent of the world's lead will be produced by this blast furnace process.

The commercial growth of the new process has, in fact, been almost as remarkable as its invention and technical development and began in 1952 when prospects were considered bright enough for steps to be taken to interest the outside world in it. In that year Neave went, with Board approval, to the U.S.A. 'with a view to interesting some American firm or firms in the Process' and the first visits from other companies to the new furnaces were in June 1953 when the Presidents of American Smelting and Refining Co. and American Zinc Lead and Smelting Co. visited Avonmouth. Neither made an offer for the process rights and, up to this day, the U.S.A. has been the one major industrial nation which has shown least interest in the process. The zinc price slump in 1953, of course, was unfavourable for sales of a new process but the price had recovered by the time that the proposed licensing arrangements were put on a sound basis in 1956. This step included the conversion of the first-ever subsidiary of National Smelting, National Processes Limited (incorporated 1927), into Imperial Smelting Processes Limited, a company which was to act as Technical Adviser and Engineering Consultants to all concerns who might become interested in the process. B. G. Perry (no relation of the R. G. Perry of sulphuric acid fame mentioned earlier in this book), the chemical engineer who had been immediately responsible for the engineering design and supervision of the construction of the furnaces completed at Avonmouth in 1950 and 1952, was made Managing Director. The story of the remarkably rapid spread of the new process round the world that followed this decision and the presenting of papers by S. W. K. Morgan to the Institution of Mining and Metallurgy and other learned bodies from 1955 onwards, is contained in a great bulk of technical literature and films which have been produced in the last ten years about the Imperial Smelting Process, as it is now called.

It is appropriate to end this account of the process by summarizing briefly its significance in the development of zinc smelting generally.

For the first time ever, zinc and lead could be produced at the same time, from the same furnace, and from mixed zinc/lead concentrates by an internally heated blast furnace instead of by externally heated retorts. As a result economies over traditional processes are effected at every stage from the mill to the casting bay. There is no need to separate mixed ores into predominantly zinc and predominantly lead concentrates in the mill for separate treatment by separate processes. Internal heating permits a far larger vessel to be used than in the case of the traditional externally heated horizontal or vertical retorts. One horizontal retort will produce only 70–80 lb. of zinc a day ($12–13\frac{1}{4}$ tons/day in a furnace containing 382 retorts), the vertical retort less than eight tons and the blast furnace currently up to 175 tons or more and economies in labour are commensurate.

The significance of this process for the British zinc industry in the post-war epoch has been that it has raised the industry far above traditional mediocrity and put it in the forefront of the metallurgical world at a time when, otherwise, it might so easily have succumbed, in an insufficiently protected market, to zinc produced by the electrolytic process overseas.

As always throughout the history of zinc smelting in Britain, both before and after the foundation of the modern zinc industry in 1917, the industry's ability to survive and flourish in the future on the basis of this great development depends on the Government's willingness to ensure for it peace-time conditions which do not place it at too great a disadvantage in the face of overseas competition. This supremely important aspect of the zinc industry will be summarized in the last chapter of this book.

CHAPTER FIFTEEN

The International Zinc Market and the Zinc Duty Question

NEED TO VIEW THE BRITISH ZINC SMELTING MONOPOLY IN A
WORLD SETTING—THE ESTABLISHMENT OF A LONDON METAL
EXCHANGE AND ITS LIMITED EFFECT ON PRODUCING AN
UNRESTRICTED FREE MARKET IN ZINC—THE DANGER OF A
COMPLETELY FREE MARKET TO INDUSTRIAL STABILITY
HOW ZINC MINES AND SMELTERS SHARE THE PROFITS
THE TREATMENT CHARGE—FIFTY YEARS OF ZINC CONCENTRATES
SUPPLY CONTRACTS—MAJOR IMPORTANCE OF THE ZINC PRICE
FORTY YEARS OF ATTEMPTS TO MAINTAIN CARTELS TO
ELIMINATE FLUCTUATIONS—OTTAWA AND PROTECTION ALTER
THE OUTLOOK—FAILURE OF PRE-1939 ATTEMPTS TO WIN
ADEQUATE PROTECTION FOR THE BRITISH ZINC SMELTING
INDUSTRY—THE 1939 AGREEMENT WITH THE EMPIRE PRODUCERS
THE WAR INTERVENES—MODIFICATION OF THE 1939
AGREEMENT IN 1950—THE EFFECTS ON THE PRICE OF DEVALUATION,
THE KOREAN WAR AND THE REOPENING OF THE
LONDON METAL EXCHANGE—GYRATIONS AGAIN
THE COMMON MARKET PRICE DIFFERENTIAL AND THE
RENEWED APPLICATION FOR INCREASED PROTECTION
REASONS WHY THE APPLICATION WAS REJECTED
THE UNITED NATIONS STUDY GROUP—A NEW BOOM AND A NEW
CARTEL WITH NEW OBJECTS—THE PRODUCER PRICE
FUTURE PROSPECTS FOR THE WORLD ZINC INDUSTRY
THE CHALLENGE BEFORE IMPERIAL SMELTING
ZINC PRICE GRAPH

IMPERIAL SMELTING CORPORATION has been the sole producer of zinc in Britain since 1933 but few have ventured in all the subsequent years to impute that it has thereby become imbued with the alleged sins of a non-nationalized monopoly. This is because, although sole British producer, it has never yet supplied more than half of Britain's zinc requirements. The remainder has come from overseas and therefore any assessment of the value of the part played by the British zinc smelting industry in the past fifty years must be made in an international and not a purely British setting.

More immediately, it will have become obvious from the many references in this book to the zinc price that this, naturally enough, has been the basic factor governing the level of prosperity of the industry since, at least, Champion's time.

It is a platitude of economic theory that, in a free market, price is governed by the interplay of supply and demand. After years of informal meetings in the Jerusalem Coffee House and elsewhere, a formal Metal Exchange was set up in London in 1882 and the price that the British zinc industry received for its zinc was determined by its transactions throughout its existence up to 1964, except for 1939–52. The intention behind the founding of the London Metal Exchange ('L.M.E.') was to provide an international free market for zinc and other non-ferrous metals but during the greater part of this Exchange's existence the conditions governing supply have been far from free and even demand has sometimes been artificially inflated to keep up the price.

However, any artificial restriction of free dealing has usually been the reverse of sinister. The main reason for it has already been emphasized in this book. A heavy industry, such as zinc mining or smelting, requires enormous and increasing capital outlay and attracts to itself hundreds or thousands of employees to produce a product which is basic and important in a modern economy. In the interest of economic and social stability this industrial superstructure should not be allowed to be endangered by speculation and by the effects of temporary gluts on the world free market price.

The simple answer to this problem, for the majority of products dealt in internationally in this imperfect world, is protection for the domestic producer by import duties. The struggle to obtain adequate protection for home produced zinc ever since Britain abandoned free trade in the early 'thirties has, of course, been an important factor in the attempt to procure a normal level of prosperity for the industry in this country. But it is not the only feature influencing zinc or, for that matter, other non-ferrous metals produced here.

A minor but, nevertheless, important feature is that price variations have also produced internal complications within the industry. The fact that the price paid for zinc metal by consumers has fluctuated with the market means also that it is not possible for the mines to sell concentrated zinc ores to the smelters at fixed and stable prices. Otherwise the smelter would gain disproportionately at the expense of the mine in times of high zinc prices while the reverse would be the case when zinc prices were low. The price that is ultimately paid by the consumer, whether it is a free or a controlled price, has therefore to recompense both the mine and the smelter for their services in producing the metal from the earth. The proportion of the price that each of the two receives is adjusted between them on the basis of a formula known as the 'returning charge' and computed on a basis now hallowed by tradition. Briefly, the 'returning charge', or 'treatment charge' as it is more accurately called, is the charge which the smelter levies from the mines on a ton of zinc concentrates to cover the cost of converting the ore into metal plus a reasonable margin for profit and metal selling expenses. This charge varies considerably both in accordance with the zinc metal price and also with fluctuations in the zinc ore market itself.

Therefore, although the dominating feature in the prosperity of both mine and smelter is the zinc price, a lesser but most important feature, affecting the prosperity of the smelter at the expense of the mine or vice versa, is the 'treatment charge'. Unlike the zinc price this is not governed basically by the interaction of world supply and demand for zinc but by supply and demand for zinc concentrates as between the mines and the unintegrated 'custom' smelters.

As more and more smelters have come into operation since the beginning of this century it has become increasingly difficult for any unintegrated smelter to escape from the operation of these economic laws and make a special arrangement with a mine, even if it is directly associated with that mine, at more than the 'free market' returning charge. This is because mines wish to become more and more assured of the most profitable outlets for their concentrates at the open market treatment charge.

The history of the contracts made since 1917 to supply The National Smelting works of Imperial Smelting Corporation with concentrated zinc ores provides a clear illustration of this trend. The background to its first contract has already been outlined in Chapter 3.

The new National Smelting Company, in its agreement with the Minister of Munitions of 11 May 1917, bound itself to take only 25,000 tons a year of the Australian concentrates for ten years from 1 January 1918, at a returning charge of £5. 10s. 0d. a ton based on zinc at £23 a ton with 3/– a ton increase for every £1 rise in the price of zinc. These and the supporting terms secured to the Company an amount equal to about 17 per cent of the metal price. This percentage is more or less in line with the treatment charge which European zinc smelters are currently receiving but the 1917 treatment charge was considered highly favourable in the 'twenties when labour and other costs were so much lower than those ruling today.

The Australian contract was undoubtedly the principal source of supply for National Smelting in those years and Australia still remains the principal source of supply although to a much smaller extent than formerly.

There were other and minor sources of supply, however, even in these early years. With the acquisition of a third of the shares of Burma Mines in 1924 Burma concentrates gradually came into use in small quantities from 1932. There is also mention of Monarch mine concentrates and Buchans River. Also, until the formation of The Consolidated Zinc Corporation in 1948 restricted the sphere of interest of Imperial Smelting almost entirely to its industrial activities in Britain, the Company, under the stimulus of W. S. Robinson, was continually looking for possible zinc deposits to mine overseas although none of its prospects ever got to the production stage.

The contractual position after 1930 is somewhat obscure owing to the fact that the contracts prevailing are not available for inspection. The statement issued by Imperial Smelting to its shareholders in 1948, just before the formation of The Consolidated Zinc Corporation, sums the story up very briefly by saying that, during and after the 1914–18 War, 'very large contracts extending over a period of ten to fourteen years were made by the British Government for purchase of zinc concentrates. These contracts expired in 1930 and were replaced by a fifteen years' sales arrangement between the Broken Hill mines, your Corporation (i.e. Imperial Smelting), and The Electrolytic Zinc Company of Australasia Limited.' From further brief details given in the Chairman of the Zinc Corporation's speech on 28 July 1930, it appears that this new arrangement of 1930 took several months to negotiate and covered the whole of the output of zinc concentrates of the Broken Hill field. There is no indication of what scale of treatment charge was included but, as the expressed intention was to carry on the Imperial policy which had inspired the previous contract and as 1930–31 saw the worst ever slump in zinc in the present century, it may be assumed that the charge is hardly likely to have been more onerous to National Smelting than the previous charge.

This new Australian contract of 1930 ran until June 1945 but, of course, as has been explained in Chapter 12, the Government took over responsibility for its implementation during the war years. When it ended, just after the conclusion of the European war, as was reported to the shareholders in 1948, 'prevailing World conditions have made it impossible to purchase supplies for longer than periods of six months. Recently, however (1948), a short-term contract was concluded with The Zinc Corporation Limited and New Broken Hill Consolidated Limited covering a part of the Corporation's requirements for the next three years. In the opinion of your Directors this must be looked upon as an interim measure only. They further consider that the direct association with producers provides the most satisfactory solution of the problem of adequately and regularly meeting the raw material requirements of your industry.'

Between 1917 and 1930, in fact, Australian zinc concentrates had changed from being a commodity wanted by few into the necessary raw material for an increasing number of smelters round the world and this was being reflected in a shrinking in the treatment charge out of which the smelters must make their profits.

After the Second World War this broadening of the market for zinc concentrates continued and, although Imperial Smelting maintained very substantial contracts for Australian concentrates, the smelter also had to negotiate with many other mines all over the world in order to supplement its basic feed of Broken Hill material. The invention and operation of the Imperial Smelting zinc blast furnace encouraged this trend since Imperial Smelting was now able to buy and treat mixed zinc/lead ores and concentrates where previously such complex materials had not been economic to mine.

Fluctuations in the treatment charge, affecting the proportion of the profits of the whole zinc production process which are assigned to mine or smelter, have indeed a very considerable influence on the prosperity of the British zinc industry but the other important factor which has an even greater influence on profits is, of course, the zinc price.

Appendix I shows the violent fluctuations that have occurred in the free market price of zinc since 1918, fluctuations which, however useful to the market speculator, have not been an encouragement to steady development on the production side of the industry.

Before 1914 neither production nor sales locally in Britain from the small smelters described earlier in this book nor output of zinc oxide concentrates from the small British mines to supply these smelters was on a scale sufficient to cause violent price fluctuations. When zinc sulphide concentrates started coming into Europe from Australia at the end of the nineteenth century they were bought up, imported, and used by a monopolistic Cartel organization consisting of the German controlled Australian Metal Company in Australia and Metallgesellschaft and one or two other big metal-producing companies in Germany and Belgium as has been described in Chapter 3. Part of these concentrates reached Britain through German control of certain smelters in the Swansea area already outlined in Chapter 2. This system offered little scope for competition and price variations.

The 1914–18 War removed the German colossus from the scene and left a world in which the U.S.A. had increased production substantially during the war and the major part of the Australian concentrates were finding their way to continental smelters via the Board of Trade contracts with National Smelting. Over-production after 1925 and post-war economic troubles had depressed the zinc price even before the crisis of 1930–31 and it is not surprising to find that the Cartel idea came back in 1928 in an attempt to stop a further fall in the zinc price. To quote again from the speech of the Chairman of The Zinc Corporation of 28 July 1930:

That some adjustment in production (i.e. of ores and concentrates) was necessary became clear in 1928 and again in 1929, and, as you were then informed, an association of producers was brought into being, which for a time promised good results. But collective action in the case of zinc is not so easy as in that of lead, where the chief producers are few in number and the sources of supply of ore or metal relatively few. In the case of zinc, not only is there a host of producers, many of them small, scattered over a large territory and drawing ore supplies from an area larger still, but there is a wide diversity of interests among them. There are big mines owning reduction works, there are mines without reduction works, and there are smelters without mines. There are customs smelters and treatment works linked up with ancillary or associated industries who look for their profits from the latter—sometimes at the expense of spelter and who speculate in metal selling against every ton of concentrates they buy. All these various parties are difficult to keep together and in the crisis at the end of the last year the European convention or Cartel quickly expired as a result of clashes of interests between customs smelters and producers, or between Continental smelters and American interests. The immediate result was chaos, producers of zinc ores and concentrates losing heavily without corresponding loss to the customs smelters. The industry is going through a period of severe adjustments which the producers (i.e. of ores) feel is very largely at their expense, for while every fall of £1 per ton in spelter reduces by 8s. the actual price received for a ton of 50 per cent zinc concentrates, delivered to the customs smelters, it does not materially affect the cost of smelting or treatment and the revenue of the smelter remains relatively constant, while that of the producer of concentrates suffers severely. When customs smelters recognise that it is the price of metal which is of chief importance to the producer (of ores), and when the producers of ores and concentrates appreciate the fact that while metals are produced nationally they are consumed internationally, and that the consumer has a very definite right to determine whether he will accept his requirements in ore or concentrates rather than metal, the road to a better regulation of the industry will be cleared.

I am happy to say that after prolonged negotiations a base for a new world Cartel has been found, and we are hopeful that it will be brought into successful operation at no distant date.

National Smelting took no traceable part in the formation of the preliminary arrangement of 1929–30 between 'producers, smelters and distributing merchants' as it was producing only about 15,000 tons of zinc a year from Swansea until early 1929. However, through Captain Oliver Lyttelton (now Lord Chandos) it took a very active part in the promotion of the 1930–34 Cartel.

W. S. Robinson reported to the Board on 16 October 1930:

A meeting of European Zinc Producers is being held in Paris on the 20th and 21st and will be attended by our representatives. Consideration will be given to the creation of a Cartel for the purpose of centralizing selling, regulating production, and obtaining complete statistics of the world's industry. The fall in the price of spelter will, it appears certain, bring about, at a not very distant date, all the required adjustments in the industry without the aid of a Cartel, but nevertheless closer association of producers of zinc appears to be justified and our attitude will be sympathetic to any sound scheme.

Six months later, on 16 April 1931, W. S. Robinson reported:

Attempts to form a Zinc Cartel in the latter half of 1930 and early this year failed largely through the hostility of the Asturienne and Norwegian Group.

but that the head of this Group had recently returned with alternate proposals.

By July 1931 the formation and objects of the International Zinc Cartel were outlined to the Board but details of the Heads of Agreement are not available. On the production side the output of each member Group appears to have been limited by reference to something called the 'Ostend basis'. For example, in October 1931 the Cartel was reported to be in a healthy position with falling stocks and production permitted at 55 per cent of the 'Ostend basis'. It appears that although it was open to a member Group to produce more than its quota it was penalized if it did so.

After a Cartel meeting on 28 April 1932 it was reported that 'since the Cartel was introduced in July 1931, the daily production of members of the Cartel has been reduced from 1,810 tons of zinc to 1,558 tons. The monthly production in March 1932 was only 48,318 tons against 56,111 tons in July 1931'. But, by August 1932, production was reduced to 45 per cent of the Ostend quotas and, as this meant a further cut in production for the 'Australian Group', the August furnace shut-down at Avonmouth and Swansea was prolonged. The Cartel were disappointed, however, that, for all their efforts, the price of zinc had continued to fall and the gold price of zinc at £8. 4s. 10d. (£10. 11s. 3d. L.M.E.)* in April 1932 was the lowest ever reached. Inevitably many producers were said to be finding difficulty in keeping going.

*Quin's Metal Handbook records a 'low' of £9 13s. 9d. L.M.E. in 1931 but this was also a 'gold price' as it was reached just before Britain abandoned the Gold Standard on 21 September 1931. The value of the currency against gold dropped by more than a quarter within a few days of that date (Oxford History of England).

A new element had appeared in the situation in March 1932 when W. S. Robinson expressed the view that 'it appears as if the new German tariff on zinc must inevitably lead to a dissolution of the existing Cartel'. The world economic crisis of 1930–31 had, in fact, been so severe that ideas of national protection were back in fashion for the first time in over half a century. Later in 1932 W. S. Robinson visited the Ottawa Conference and from the conference emerged Imperial Preference and the U.K. Import Duties Act of 1932. Cartels hitherto had been founded on the basis of free trade without allowing for the distortions that national tariffs would introduce into metal prices. So this 1930–34 Cartel, of which Captain Oliver Lyttelton of Imperial Smelting was by now Chairman, although it was able to survive through persuasion, renewals and attempts to frame a 'permanent scheme', for a few months at a time until 31 December 1934, was then suspended indefinitely except perhaps for its useful statistical services which members hoped would continue.

The reasons for its decease were contained in a letter from three of its leading personalities to its President which is worth quoting in full:

THE BRITISH METAL CORPORATION LIMITED

London, E.C.2.

9th November, 1934.

Monsieur Saint-Paul de Sincay,

Entente Internationale des Producteurs de Zinc,

23, rue Belliard,

BRUSSELS.

Dear Sir,

We recall to your attention that we were appointed as a Committee on July 16th to examine the basis for a new Zinc Cartel.

Today it is clear to all concerned that the London Metal Exchange quotations do not reflect the working of the law of supply and demand for Spelter, and it is clearly a condition precedent to the formation of a new Cartel that this question should be resolved. The reason why this situation has arisen is because the British Empire producers have been obligated to sell their metal at the 'world price'. The 'world price' has been taken to mean the London Metal Exchange quotation, which, under existing circumstances, is a quotation for metal other than British and which is therefore subject to duty when entering the United Kingdom.

This has the effect that the sale of a small tonnage of foreign Spelter on the London Metal Exchange, for which there can be no British buyer other than a speculator, depresses the price for the whole world outside the U.S.A. and leads to the abrogation of the normal functions of the Metal Exchange.

It is clear, therefore, that this Committee cannot make substantive proposals for the future regulation of the industry until the above problem has been resolved, but we have reason to believe that these questions are now under discussion and that a definite conclusion is unlikely to be long delayed.

In the meanwhile, we suggest that the present Cartel at the date of its expiration suspends all its functions except its statistical services.

In our opinion it will not be difficult to formulate sound proposals for a new Cartel at a later date.

Yours faithfully,

(s) R. MERTON (s) O. LYTTELTON

(s) V. MIKOLAJCZAK

Put briefly, tariff disparities were upsetting the price to the disadvantage of the British and Empire producers. As will already be obvious from Chapter 8 on the 'thirties, this import duties question was a vital one for the British zinc industry and had already been artificially depressing its profitability for over two years when the Cartel came to an end. The inadequate arrangements made for the protection of the price paid for zinc to the British zinc smelting industry in 1932–35 gave it a poor start in an increasingly protectionist world, and remains one of the two main sources of its poor profitability to the present day. The approaches made to the British Government to remedy this grievance and the validity of the reasons why they have hitherto been rejected will be discussed later in this chapter as it is still possible that the nation's zinc industry may founder through the failure of the Government to recognize the seriousness of this issue. Admittedly a compromise agreement was reached with the Government in 1939 which would have helped the industry to make a profit if the conditions of the previous seven years had continued. However, war broke out soon afterwards, before the industry could derive any benefit from the compromise, and the whole issue had to be reopened in altered circumstances after the war. It has not yet been solved.

The negotiations of the 'thirties, however, provide an interesting commentary on the conflicting influence of past and future on the proposals put forward.

Before 1932 zinc metal was imported duty free in the aftermath of the liberal free trade epoch. Under the Import Duties Act which came into operation on 1 March 1932 imports of foreign zinc metal were made subject to the general *ad valorem* duty of 10 per cent while, with the sun setting on the British Empire, imports of zinc metal from the Dominions, India and Southern

Rhodesia remained free of duty. Imports of zinc concentrates from all sources, being classed as raw material, remained and always have remained duty free. The detailed case which Imperial Smelting addressed to the Import Duties Advisory Committee on 10 March 1932, presumably with the forthcoming Ottawa Conference in mind, stresses the importance of, at least, maintaining this level of duties. The several reasons given included the debased currencies of certain foreign countries (e.g. post-war Germany) which had lowered their costs, the higher standard of wages and higher taxation charges in Britain, and the importance of increasing employment in the smelting works and coalmines in an age of gigantic unemployment. The additional bait that was dangled was the proposal to increase domestic production of zinc from the 1932 level of 15,000 tons to 60,000 tons a year if satisfactory tariff protection were obtained. This figure of 60,000 tons was later to assume great significance in 1939.

The 1932 Ottawa Agreements provided for the continuation of the general *ad valorem* duty of 10 per cent on foreign (as opposed to Empire produced) imports of zinc for as long as Empire producers met British consumers' requirements at world prices (later L.M.E. prices).

W. S. Robinson says that, at the Conference he left the British fabricators to argue things out with the Empire producers and that the fabricators (i.e. manufacturers) dictated this decision i.e. their price for agreeing to keep cheap foreign zinc out was that Empire producers should supply at the lowest possible 'world' prices.

It is important, therefore, to consider what the 'world' production position was in 1932. Foreign producers meant mainly the U.S.A., German*, and Belgian producers. 'Empire' producers meant primarily the Canadian producers, Consolidated Mining and Smelting (now Cominco), and Hudson Bay Mining and Smelting, as very little Australian or Rhodesian zinc was being imported into Britain at this time. The Canadian zinc industry had grown up since the 1914–18 War using Canadian concentrates and cheap hydro-electric power which made possible the use of the electrolytic process for zinc production. This Canadian industry, however, was no part of the original 'Empire scheme' as formulated in the 1914–18 War. That scheme, it will be remembered, was based entirely on the idea of disposing of Australian zinc concentrates by setting up smelters in Australia, Britain and other parts of the Empire. The idea of a Canadian industry using its own concentrates would probably have been very unpopular with those who framed the scheme.

---

*An extant report from W. S. Robinson on his visit to Germany in 1938 shows how very rapidly the German zinc industry recovered—almost to its pre-1914 stature—under the Nazis.

By the 'thirties, however, the Canadian industry had become a power to be reckoned with, particularly as it was using the world's cheapest (in running costs) and most modern zinc producing process at that time, the electrolytic process, and the history of the past thirty years of the British zinc industry has been dominated by its relations with the Canadian industry. The local British industry at present produces just over a third in tonnage of the country's requirements although it meets about 80-90 per cent of the country's demand for the lower grades of zinc, and the bulk of the balance comes from Canada, principally to meet the demand for the higher purity grades.

The basis of the existing situation was laid by the Ottawa Agreements of 1932 and the position has, by accident rather than design, been adjusted even more in favour of the Empire and foreign producers by several decisions since that date.

In December 1934 the Board found that the Ottawa system was already far from improving the dismal position of the British zinc industry and applied for a minimum operative duty of £2 a ton of zinc. This figure was based on the assessed advantage in cost of production possessed by leading continental competitors owing to their lower wages, transport and taxation charges.

Almost simultaneously with the application and with the termination of the Cartel on 31 December 1934, the Import Duties Advisory Committee were asked by the Board of Trade to examine the working of the Ottawa Agreements in connection with Dominion lead and zinc imports. The report submitted by the Committee in August 1935 represented an effort to establish a single formula to meet both the difficulties arising out of the Ottawa Agreements and the requirements of the British zinc smelting industry.

As a result of the report, the duty on foreign zinc was reduced to 12/6 a ton which was considerably below the former 10 per cent *ad valorem* duty. At the same time the Empire producers were relieved of their obligations to meet British customers' requirements at world prices.

The object of this adjustment was to clear the ground for yet another international agreement to regulate the production and consumption of zinc. Imperial Smelting agreed to give the Advisory Committee's suggestions a fair trial 'after lengthy discussion and protests' and after very definite assurances from the Committee that they would not hesitate to take any steps necessary to protect the industry from external competition.

In the Hitler-Mussolini ridden atmosphere of late 1935 the pious intention to form another International Cartel was doomed to failure from the beginning, although Captain Oliver Lyttelton did everything possible to bring such a Cartel into being as was reported to the Board in 1936. Meanwhile, naturally enough, Imperial Smelting found it quite impossible to operate profitably under the weakened tariff arrangements of 1935. Accordingly representations were made to the Import Duties Advisory Committee in 1937 and 1938 to restore a high rate of duty on foreign metal. The Committee, however, considered that this step by itself would be insufficient in view of the large volume of imports from Empire sources and opened negotiations with the Dominions. These eventually led the Treasury to raise the import duty on foreign zinc to 30/– a ton with effect from 26 May 1939 as has been mentioned in Chapter 8. The figure of 30/– was chosen as being practically equivalent to 10 per cent of the average import value of zinc at the time. This 30/– has since become a tradition which it has proved extremely difficult to alter.*

To this step the agreement between Imperial Smelting and the Empire producers of the same date was complementary. The keystone of this latter agreement on Imperial Smelting's side was their consent to limit production of virgin zinc to 60,000 tons in any calendar year. In return the Empire producers agreed to pay Imperial Smelting, out of the increased price that they would receive for their metal, a subvention of 10/– a ton (decreasing to a minimum of 1/3 with the zinc price rising above £18) on every ton of Empire metal imported and a contribution towards a rebate of 17/6 a ton of the zinc content of products exported by fabricators from Britain.

Although detailed figures are available it would be superfluous in a book of this nature to give anything more than a brief outline of how these arrangements worked out and the effect that they had on the prosperity of the British zinc industry and, more particularly, on the relationship between the Canadian and British producers.

The 1939–45 World War broke out only three months after they were made and although, as has been seen in Chapter 10, the British zinc industry endeavoured, in spite of air raids and labour shortages, to increase zinc production much greater reliance had to be placed on Canadian supplies than was ever envisaged when the arrangements were made. In particular, through inadequate capacity on the vertical retorts and refluxers to provide enough zinc for refluxing into alloy grade metal for munitions, the country came to depend almost entirely on Canadian metal for zinc alloy production.

In spite of increased Canadian imports, British zinc production had also to be increased, with active Government encouragement, to well over the 60,000 tons level during the war and, in the circumstances of the time, the Canadians naturally did not object.

---

*30/– is, of course, only just over 2 per cent of £65, which is considered as about the minimum marginal price for zinc in the context of present day production, wages and other costs.

The 60,000 tons limitation was, in fact, virtually never observed at all. The Canadians raised this question again in 1947 and, while not objecting to the excess production in the circumstances of post-war shortage then existing, reserved their right to object if conditions changed. They further asked that consideration should be given to suspending the subvention while Imperial Smelting's production exceeded the permitted tonnage. Discussions on the whole 1939 agreement followed and the eventual conclusion reached was that, while the existing duty of 30/– a ton might, at prevailing values, be wholly inadequate either as protection to domestic producers or as preference to Dominion producers, it would be wiser to retain it as establishing a principle than remove it and be faced in the future with an appeal for its reinstatement. The minimum subvention of 1/3 and the export rebate to fabricators of 17/6 were also allowed to stand. The question of the British industry observing the 60,000 tons limit appears to have been left in the air.

Eventually, in 1950, the Imperial Smelting Board asked A. M. Baer, who had recently joined it, to take up again the whole question of the Empire producers. In view of Imperial Smelting's expectation that in succeeding years zinc production would be greatly expanded through the promise of the new zinc blast furnace, an undertaking was formally entered into with the Commonwealth producers that the 60,000 tons limit should be abolished in consideration of the withdrawal of the subvention payment by those producers.

Great changes had of course already taken place in the zinc marketing position, first through the devaluation of the Pound in September 1949, when the official price was raised from £63. 10s. 0d. to £87. 10s. 0d. subsequently reaching a new high during that year of £106. With the outbreak of the Korean War in June 1950, further severe pressure was put on world prices of zinc and the Ministry raised their official quotation to £151. When the shortage of zinc reached its peak in 1951, the Ministry price was raised further to £190 a ton, which price remained in force until May 1952. Later in 1952 the price of zinc began to slide rapidly and, with the reopening of the London Metal Exchange in January 1953 and the Korean War drawing to a close, the price of all commodities fell sharply and zinc was no exception. In fact, before the year ended a low point of £63. 7s. 6d. was reached.

Ever since 1949 the Imperial Smelting Board had been debating the pros and cons of making a renewed approach to the Ministry for an upward revision in zinc duty. However, following upon devaluation of the Pound and the start of the Korean War, zinc prices rose so sharply that it was felt inopportune to make this approach. The Company during these years was operating at a high rate of profit and any increase in profits would have been nullified by the Excess

Profits Duty. At the same time the case would have been strongly resisted by the British manufacturers and likewise, as a matter of principle, by the Commonwealth producers. Therefore no action was taken and Imperial Smelting was compelled to carry on through the subsequent years at a disadvantage compared to the majority of overseas smelters.

In the years following the reopening of the London Metal Exchange, gyrations on the zinc market continued with extremes of £105. 10s. 0d. and £63. 6s. 3d. At the lower levels the operation of the horizontal retorts was severely jeopardized and every effort was made to bring the new zinc blast furnace into production at the earliest possible date to counteract the difficulties then being experienced.

A new situation arose, however, in 1957 when negotiations for the establishment of the European Economic Community first began. A programme was put forward which envisaged the ultimate erection of a common external tariff on zinc metal for the six countries concerned and the abolition in due course of internal tariffs. As the existing tariffs of these countries were at different levels the process was expected to be slow so that no serious damage would be done to the more highly protected countries. As a result it was agreed that a common external tariff should eventually be established at approximately the mean of the existing tariffs, i.e. £4. 16s. 0d. a ton. These proposals were of great concern to Imperial Smelting since they would result in all European smelters holding a great advantage over the British smelter in buying their ores and concentrates in the world market.

It was, in fact, considered vital by Imperial Smelting that, if there were to be a Common Market tariff wall of this magnitude, the British zinc industry should be inside it and this view was expressed to the President of the Board of Trade by letter on 4 July 1962, and in a subsequent meeting, in a preliminary request for an increase in tariff.

When, therefore, General de Gaulle's speech of 29 January 1963 brought an abrupt end to the negotiations over Britain's application to enter the Common Market, Imperial Smelting at once put in an application to the British Government for an increase of the import duty on foreign zinc coming into Britain to the level of the Common Market external tariff, with maintenance of the existing 30/– preference in favour of Commonwealth producers, who would thus have to pay £3. 6s. 0d. a ton import duty, instead of nil as hitherto, if duty were established at the £4. 16s. 0d. level of the Common Market.

After much correspondence and a good many interviews with Board of Trade officials this approach was officially rejected on 2 July 1964, in a letter which set out obligations to the Commonwealth, G.A.T.T., and the forth-coming 'Kennedy Round' negotiations for tariff reductions as the main reasons for the Government's refusal.

The British Government, in fact, found themselves in a difficult position *vis-à-vis* Commonwealth producers and their respective Governments. These Governments had been fighting hard to protect their own interests in the event of Britain's entry into the E.E.C. and, when such entry was refused, the Commonwealth producers were in an even stronger position to resist Imperial Smelting's application.

It is worth recording that when the British fabricators were officially informed of the tariff proposals, they also showed collective resistance which proved a great disappointment to those who were seeking to establish British smelting operations on a firm and expanding basis for the long term. Although the Commonwealth producers had been supplying the industry with a fair propor-tion of its requirements over many years, it would seem that the fabricators' opposition was shortsighted in as much as the extra price payable for zinc under the new duty would only have amounted to about 2·3 per cent. In view of the possible danger of interruption to overseas supplies, Imperial Smelting were justified in thinking that this was a small premium for the industry to pay. The future, therefore, of the tariff question must presumably rest with the possible entry of Britain into the E.E.C. at some later date, although further approaches have been made to the Government since the official rejection of 2 July 1964.

During the latter years of the 1950's, the difficulties with which Imperial Smelting was having to contend were shared to some extent by the mines and smelters of overseas countries. In fact the situation in 1957 became so acute that the many Governments concerned formed a United Nations Study Group to consider possible remedies. Some Governments had already taken independent action, e.g. the United States' Government, where import quotas were instituted in addition to the existing tariff protection, as a result of which imports from other producing countries into the U.S.A. were severely curtailed. The effect of this measure was that the surplus of production, instead of being spread throughout the consuming world, was centred mainly upon the European market and, in particular, on the London Metal Exchange.

Although some recovery in price took place between 1959 and 1961, by the middle of 1962 zinc had once again relapsed to around £63 per ton. However, during 1963 and 1964 a new era of prosperity developed in the industrial countries and this, coupled with the entry of the United States into the Vietnam War, caused a fresh scramble for metal supplies. The position reached an acute phase in the summer of 1964 when the London Metal Exchange price for zinc became almost out of control. Following a meeting of leading producers in July of that year, under the leadership of Imperial Smelting, a producer price of £125 per ton was agreed and adhered to by all the leading smelters to prevent a run-away market and to avoid the long-term dangers of the substitution of zinc by other relatively cheaper materials. This move by the producers, assisted by releases of zinc from the U.S. stock pile, had the desired effect and sales of zinc were maintained on an orderly basis.

The availability of new mine production to smelters over the past few years and continued releases from the U.S. stock pile have made possible three further reductions in the producers' price which now stands at £98. It is believed that this whole transaction is the first occasion on which zinc producers have agreed to a commodity price control arrangement designed to keep prices at a lower level than the open market forces dictate.

This review of the past fifty years has attempted to portray the difficulties, due to inordinate price fluctuations, which long-term planning has had to face when the problems of opening new zinc mines and of the erection of new zinc smelters have had to be considered. There is little doubt, however, that much as the producers may try to stabilize the price of zinc, the real difficulties with which they have to contend are the tremendous swings in the world's economy which have taken place so frequently over the past fifty years and which are likely to continue. However, those entrusted with the responsibilities of planning large mining and smelting operations must take an optimistic view of the long-term economic progress of the world. Although there seems little hope of smoothing out the short-term slumps and booms, the evidence of the past proves that consumption of a basic metal which does not out-price itself against other commodities is bound to expand. If, therefore, we look forward over the next ten years and dare to anticipate an average 3 per cent annual increase in consumption, the industry will be faced with the problem of raising zinc metal production in the free world by around 40 per cent. With consumption at present running at around $3\frac{1}{2}$ million tons a year, extra production of nearly $1\frac{1}{2}$ million tons may be required by the end of this period. Thus, despite the immediate threats of greatly increased mine production over the next few years, it is satisfying to prophesy that all this and more will be necessary if these estimates are proved correct.

Hence there must be new smelters and extensions to existing smelters and the challenge to Imperial Smelting will be to improve still further the technique and performance of the zinc blast furnace to ensure that the world gains the maximum possible advantage from the manifest ability of this British invention to produce economically an increasing range of non-ferrous metals from a wide range of ores.

The modern British zinc smelting industry has proved itself capable of rising to meet the problems of change, depression, and national emergency over the past fifty years and, provided that its importance as a vital basic industry and progenitor of basic industries and the problems that it faces in the world of the day are recognized by Governments and the nation in future years, it will continue to play the part that it was intended to play when it was created in 1917.

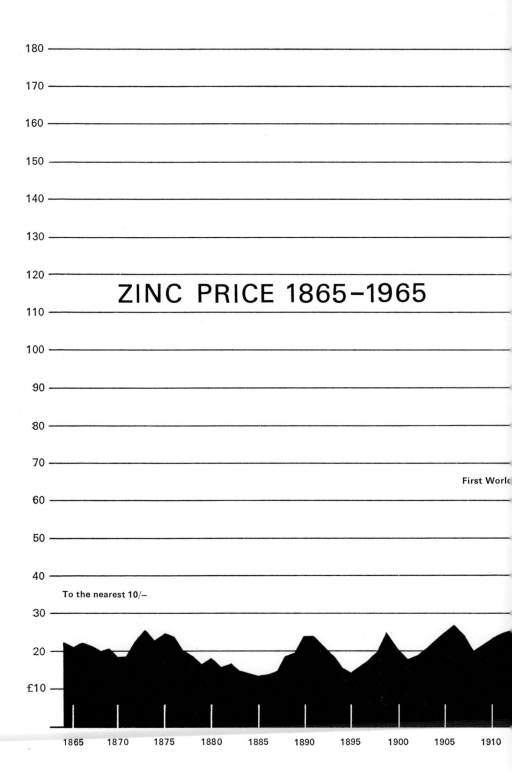

ZINC PRICE 1865–1965

To the nearest 10/–

First Worl[d]

| 180 |
| 170 |
| 160 |
| 150 |
| 140 |
| 130 |
| 120 |
| 110 |
| 100 |
| 90 |
| 80 |
| 70 |
| 60 |
| 50 |
| 40 |
| 30 |
| 20 |
| £10 |

1865   1870   1875   1880   1885   1890   1895   1900   1905   1910

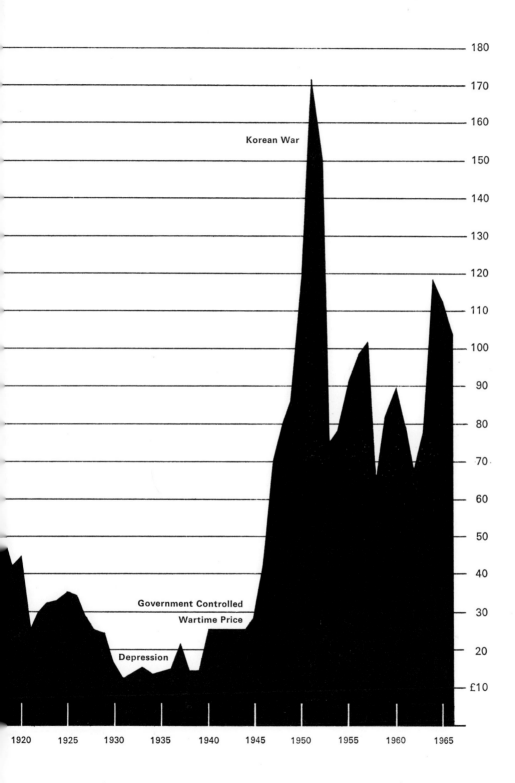

180

170

Korean War

160

150

140

130

120

110

100

90

80

70

60

50

40

Government Controlled

30

Wartime Price

20

Depression

£10

1920   1925   1930   1935   1940   1945   1950   1955   1960   1965

APPENDIX I

The Zinc Price from 1859

(Annual Average Price)

	£	s.	d.		£	s.	d.		£	s.	d.
1859	21	0	0	1896	16	11	10	1932	13	13	10
1860	20	10	0	1897	17	9	10	1933	15	14	11
1861	17	18	0	1898	20	8	9	1934	13	15	2
1862	18	6	0	1899	24	17	2	1935	14	3	6
1863	18	2	0					1936	15	0	8
1864	22	2	0	1900	20	5	6	1937	22	6	11
1865	20	12	0	1901	17	0	7	1938	14	1	7
1866	21	18	0	1902	18	10	11	1939	13	19	6/
1867	21	0	0	1903	20	19	5		14	4	3
1868	20	4	0	1904	22	11	10				
1869	20	8	0	1905	25	7	7	1940	25	15	0
				1906	27	1	5	1941	25	15	0
1870	18	10	0	1907	23	16	9	1942	25	15	0
1871	18	8	0	1908	20	3	5	1943	25	15	0
1872	22	8	0	1909	22	3	0	1944	25	15	0
1873	26	3	6					1945	28	16	5
1874	22	17	7	1910	23	0	0	1946	43	2	8
1875	24	1	4	1911	25	3	2	1947	70	0	0
1876	23	6	3	1912	26	3	4	1948	80	0	6
1877	19	18	8	1913	22	14	3	1949	87	10	0
1878	17	17	10	1914	23	6	8				
1879	16	12	0	1915	66	13	8				
				1916	68	8	11	1950	119	8	8
1880	18	7	1	1917	52	3	6	1951	171	16	3
1881	16	5	6	1918	52	3	11	1952	149	8	3
1882	16	19	9	1919	42	5	3	1953	74	14	0
1883	15	6	6					1954	78	5	4
1884	14	8	11	1920	45	4	5	1955	90	13	4
1885	13	19	11	1921	26	4	0	1956	97	14	3
1886	14	5	1	1922	29	14	2	1957	81	11	7
1887	15	4	0	1923	32	18	6	1958	65	18	0
1888	18	1	9	1924	33	12	0	1959	82	4	8
1889	19	15	7	1925	36	5	0				
				1926	34	2	8	1960	89	5	11
1890	23	5	0	1927	28	9	11	1961	77	14	7
1891	23	5	1	1928	25	5	4	1962	67	9	2
1892	20	16	7	1929	24	17	7	1963	76	14	1
1893	17	18	1					1964	118	2	6
1894	15	8	7	1930	16	16	9	1965	113	3	2 (L.M.E.)
1895	14	12	2	1931	12	8	10	1966	102	16	8 (L.M.E.)

NOTES

(a) Figures from 1859–72 inc. from Report of the Departmental Committee of Board of Trade on Non-Ferrous Mining Industry dated 17 March 1920.

(b) Prices from 1873 are quoted by kind permission of the Metal Bulletin.

Production and Tonnages of the Ir

Year	Zinc (tons)	H$_2$SO$_4$ 100% (tons)	Zinc Dust (tons)	Solid Aluminium Sulphate (tons)	Cupri (gal
1929	23,075	49,000	—	—	—
1930	19,780	54,130	—	—	—
1931	14,889	58,802	—	—	—
1932	23,799	70,414	37	—	—
1933	38,911	69,974	97	—	—
1934	46,513	74,224	211	5,711	—
1935	64,819	91,657	627	12,910	66,7
1936	65,083	132,371	755	10,962	101,4
1937	66,004	140,127	1,365	12,785	91,4
1938	55,835	126,787	873	11,384	106,1
1939	50,967	148,184	1,965	15,704	236,9
1940	61,921	142,244	4,894	14,316	446,4
1941	68,679	124,373	4,973	3,035	310,6
1942	73,001	160,365	5,362	Blitzed	207,5
1943	69,692	164,596	8,539	—	164,6
1944	72,758	164,099	7,234	—	294,8
1945	62,454	165,587	4,056	—	252,7
1946	66,794	163,985	7,916	—	362,6
1947	68,856	184,585	6,812	—	257,1
1948	73,514	201,425	6,721	12,243	271,3
1949	65,397	213,608	5,586	14,348	287,8
1950	72,176	223,475	5,849	16,206	367,9
1951	72,106	222,121	6,507	20,007	461,0
1952	68,033	213,398	4,562	14,210	503,9
1953	74,075	239,000	6,292	14,349	414,4
1954	81,263	264,044	7,241	16,896	448,6
1955	81,001	252,720	7,409	18,457	406,4
1956	81,057	253,429	6,640	16,779	379,8
1957	76,243	251,779	7,470	15,262	407,4
1958	73,826	220,775	5,990	15,448	322,9
1959	73,001	215,250	6,762	16,523	(Plant
1960	73,744	234,689	7,930	17,208	—
1961	92,362	213,557	6,844	14,264	—
1962	97,246	214,197	6,059	12,364	—
1963	99,139	217,470	7,001	10,312	—
1964	109,770	240,703	6,931	10,292	—
1965	104,956	245,694	6,912	10,372	—

*six months
†based on unsupported Works statement only.

Smelting's Main Products 1929–65

...zak oys ns)	Cadmium (lb.)	Aluminium Fluoride (tons)	Hydrofluoric Acid (as 100%) (tons)	28%/30% Lithopone (tons)	Zinc Oxide excluding Vidox (tons)
—	—	—	—	—	—
—	—	—	—	16,596	—
—	—	—	—	13,728	—
—	—	—	—	19,478	7,255
307	—	—	—	25,168	9,778
993*	—	—	—	30,651	11,038
186	—	—	—	30,772	10,347
194	—	—	—	34,021	10,983
792	140,022	—	—	35,284	12,150
623	275,968	—	—	30,781	11,572
915	233,227	—	—	36,043	10,261
150	402,473	—	—	36,642	16,523
998	328,802	93	81†	36,569	11,200
128	343,250	557	489†	33,054	8,886
614	388,686	720	532†	27,890	8,598
539	468,274	767	574†	29,292	11,859
073	391,507	785	577†	31,446	13,614
756	238,075	660	487†	37,020	19,142
884	217,685	795	632†	32,641	16,842
676	258,104	939	795†	36,835	16,843
997	221,820	1,104	1,421	37,705	12,466
783	267,910	759	1,521	41,191	13,997
330	323,109	1,007	1,625	43,186	9,356
750	336,756	1,174	1,640	17,852	6,741
527	345,933	1,128	1,649	22,831	8,572
189	302,489	762	1,616	24,257	8,337
615	311,360	992	1,690	24,445	9,278
829	241,920	1,018	2,009	21,737	8,106
925	228,480	—	1,526	20,133	9,704
741	226,240	—	1,554	17,323	8,805
964	271,040	—	2,103	20,334	7,654
380	248,640	—	2,464	18,588	7,848
207	217,280	—	2,715	17,719	6,232
235	199,360	—	2,467	16,282	5,990
452	237,440	—	2,853	14,077	6,639
528	427,840	—	3,744	—	6,727
357	510,720	—	3,645	—	6,994

APPENDIX III

The Story of Burma Mines

Mr. C. T. Fry writes:

'It was late in the 19th century that the attention of Europeans was drawn by Burmese to the enormous slag dumps at Namtu, and in 1891, Mr. A. C. Martin, of Rangoon, visited the site and took some samples. With Mr. J. Sarkies, also of Rangoon, he formed a partnership, and applied for a lease of four square miles. But for several years nothing was done until the two pioneers, along with Captain M. E. Kindersley and Mr. Maitland Kindersley, formed the Burma Mines Development and Agency Co., of which the Burma Mines Railway and Smelting Co., was a branch.

'In 1904, the Company was reorganized as the Great Eastern Mining Co., and Lord Herschell became identified with the Company. This Company sold its rights to the Burma Mines Railway and Smelting Co. in 1906, and in 1908 the name was changed to Burma Mines Ltd. At that time there was no intention to work the Mines, but only to build a railroad and smelt the Chinese slag. Forty-five miles of two feet narrow gauge railroad was completed to connect with the Government lines, and a smelter was built at Mandalay, 169 miles away.

'Smelting actually started in 1909, but no money was made as the railroad haul was too great. In 1911, a new smelter was built at Namtu on the narrow gauge line only twelve miles from the mine. After some 200,000 to 300,000 tons of slag had been mined the engineers commenced to consider that the Chinamen must have left some ore, so they began prospecting by cleaning out the old workings.

'From then on various reconstructions took place resulting in the formation in 1920 of Burma Corporation Limited (incorporated in India) which took over the previous company Burma Mines Ltd, of which Mr. Tilden Smith had been a Director since 1908 and in which he had a very large interest (in company with H. C. Hoover later President of the U.S.) which formed the basis of his massive holding in Burma Corporation Ltd., later acquired by N.S.C. at the end of 1923.'

According to W. S. Robinson, H. C. Hoover, a mining engineer, had put American money into the mine when it was reorganized as the Great Eastern Mining Company in 1904 and the mine had gone into production by 1916. It paid its first dividend in 1923.

P. E. Marmion had moved from The Swansea Vale Spelter Company to become Manager of the Mine in 1922 and he was holding that post when W. S. Robinson first inspected the mine with him in 1923. W. S. Robinson claims that Burma Mines also had blast furnaces in operation on the site and that from 1923 until overrun by the Japanese in 1941 this enterprise paid out over £12 million in dividends to shareholders and redeemed £1,100,000 of Debentures.

LIST OF SOURCES

The Board papers, Legal Documents, Annual Returns to the Registrar of Companies and files, where available of:
English Crown Spelter
The Swansea Vale Spelter Company
The Zinc Corporation
The British Metal Corporation
The National Smelting Company
National Processes
Imperial Smelting Corporation
Non-Ferrous Metal Products
Orr's Zinc White
Fricker's Metal Co., and Fricker's Metal and Chemical Co.
The Northern Smelting & Chemical Co.
Improved Metallurgy
The New Delaville Company and the Delaville Spelter Company
Swansea Vale Garden Village
National Alloys
Magnesium Holdings
Calloy

GENERAL REFERENCES
J. M. Dawkins: *Some Notes on the History of Zinc* (Z.D.A.). *Quin's Metal Handbook.* *The Metal Bulletin.* *The Mining Journal.* A. J. P. Taylor: *English History 1914–1945.* O.U.P. (1965). David Thomson: *England in the Twentieth Century (1914–1963).* Pelican (1963). Henry Pelling: *A History of British Trade Unionism.* Pelican (1963). Hansard *The Times.*

FOR THE PRE-1939 PERIOD
GERMANY
Leopold von Weise: *Beiträge zur Geschichte der wirtshaftlichen Entwichlung der Rohzink-fabrikation.* Merker Frankfurt (1902). W. F. Bruck: *Social and Economic History of Germany from William II to Hitler (1888–1938).* O.U.P. (1938). W. H. Dawson: *Industrial Germany.* Collins (1913). A. Rosenberg: *Imperial Germany 1871–1918.* Beacon Press (1964). Miscellaneous short papers issued by Metallgesellschaft. Hans Achinger: *Wilhelm Merton in seiner Zeit.* Waldemar Kramer (1967).

AUSTRALIA
There are a number of books on the Broken Hill industry, the most recently published being:
George Farwell: *Down Argent Street.* F. H. Johnston Publishing Co. Sydney (1948). *The Zinc Corporation Ltd., and New Broken Hill Consolidated Limited.* Company publication (1948). O. H. Woodward: *A review of the Broken Hill Lead-Silver-Zinc Industry.* (1952). *The First Fifty Years—A History of the Zinc Corporation Ltd.* Company publication (1956). Alfred Heintz: *The Fabulous Hill—a comprehensive review of the Broken Hill Metal Industry* (1958).

GENERAL (PRE-1939)

The History of the Ministry of Munitions in the 1914–18 War. Restricted Government Publication (1921–22). Interim Report on Certain Essential Industries 16.3.17. Cd. 9032 of 1918. Report on the Post-War Position of the Sulphuric Acid and Fertilizer Trades dated 18.2.18. Cd. 8994 of 1918. Final Report of the Balfour Committee on Industrial Policy after the War 3.12.17. Cd. 9035 of 1918. Report of the Departmental Committee of the B.O.T. on the Non-Ferrous Mining Industry dated 17.3.20. Cd. 652 of 1920. Commerce and Industry—Statistical Tables. Ed. W. Page. Constable (1919). Rt. Hon. Christopher Addison: *Politics from Within 1911–18.* Herbert Jenkins. Rt. Hon. Christopher Addison: *Four and a half Years.* Herbert Jenkins. Trevor Wilson: *The Downfall of the Liberal Party.* Collins (1966). Lord Chandos: *Memoirs.* Bodley Head (1962).

(There are, of course, many other excellent books on various aspects of the world between 1914 and 1939 but only those which refer directly to persons connected with the zinc industry have been listed here.)

TECHNICAL

T. E. Lones: *Zinc and its Alloys* (1919). *Zinc—The Science and Technology of the Metal, its Alloys and Compounds.* Ed. C. H. Mathewson. Reinhold Publishing Corp. (U.S.A.) (1959). Tomlinsons Cyclopaedia Vol. I and II (1852). Report of the Mineral Development Committee. Cmd. 7732 July 1947. H.M.S.O. Stanley Robson: *The Roasting of Zinc Concentrates in Great Britain.* Chemical Engineering Group Proceedings Vol. 11 & 12. 1929 and 1930. H. F. Brauer and W. M. Pierce: *The Effect of Impurities on the Oxidation and Swelling of Zinc-Aluminium Alloys.* Inst. Met. Div. A.I.M.M.E. Technical Records of Explosives Supply 1915–18. Manufacture of Sulphuric Acid by The Contact Process. Ministry of Munitions and Department of Scientific and Industrial Research. H.M.S.O. 1921. Beevis & Bradley: *Protective Coatings for Metals.* Reinhold Publishing Co. J. C. Carr and W. Taplin: *History of the British Steel Industry.* Blackwell, 1962. T. B. Gyles: *Horizontal Retort Practice of the National Smelting Company Limited, Avonmouth, England.* Trans. AIME, Vol. 121 (1936). W. O. Alexander: *A Brief Review of the Development of the Copper, Zinc and Brass Industries of Great Britain from A.D. 1500 to 1900.* Murex Review, Vol. I, No. 15 (1955). Murex Limited, Rainham, Essex. J. W. Gough: *The Mines of Mendip.* David & Charles, Newton Abbot (1967).

TECHNICAL
THE CHEMICAL INDUSTRY

T. I. Williams: *The Chemical Industry.* Pelican (1953). D. W. F. Hardie and J. Davidson Pratt: *A History of the Modern British Chemical Industry.* Pergamon Press (1966). *Orr's Zinc White 1898–1948* (1948). Sir Alfred Robbins: *Orr's Zinc White—A Romance of Modern Industry* (1927). L. J. Barley: *B.T.P.—Some Pre-History* (Private document). Documents of the District Court of the United States of 1945 (Civil No. 26–258) and of the Supreme Court of the United States (Appeals Nos. 89 and 90 of 1946) in U.S.A. v. National Lead Company, Titan Company Inc., E.I. de Pont de Nemours and Company.

List of Illustrations

Index